PURDEY'S

The Guns and the Family

PURDEY'S

The Guns and the Family

Richard Beaumont

DAVID & CHARLES
Newton Abbot   London   North Pomfret (Vt)

The end papers show the original wallpaper in The Long Room

*I dedicate this book to the memory of my uncle,
Hugh Sherwood, who did so much to keep Purdey's
in existence during the difficult post-war period*

British Library Cataloguing in Publication Data
Beaumont, Richard
　Purdey's: the guns and the family.
　1. James Purdey & Sons—History
　2. London (England)—Industries—History
　I. Title
　338.7'6834'00942132　　HD9743.G8J3

　ISBN 0–7153–8624–7

Colour photography by Tony Hutchings

© Text: Richard Beaumont 1984
© Colour and black-and-white photographs, and letters:
　James Purdey & Sons Ltd 1984

All rights reserved. No part of this
publication may be reproduced, stored in
a retrieval system, or transmitted, in
any form or by any means, electronic,
mechanical, photocopying, recording or
otherwise, without the prior permission
of David & Charles (Publishers) Limited

Typeset by ABM Typographics Ltd., Hull
and printed in Great Britain
by Butler and Tanner Ltd., Frome
for David & Charles (Publishers) Limited
Brunel House　Newton Abbot　Devon

Published in the United States of America
by David & Charles Inc
North Pomfret　Vermont 05053　USA

CONTENTS

Prologue	6
Family Tree	8
The First James Purdey	11
Golden Years and New Premises	35
Late Victorian and Edwardian Splendour	69
World War 1914–18	112
The Twenty-year Peace	132
World War Again	175
Masters of Their Craft	190
To the Present Day	204
Appendices	
1 Deed of Agreement changing the name of the firm to Purdey and Sons in 1877	237
2 Alterations to the Woodward/Purdey over-and-under gun after 1948	241
3 The numbering of Purdey guns 1814–1983	243
Acknowledgements	244
Index	245

PROLOGUE

On hearing the name Purdey, most people think in terms of guns and sport; they very rarely consider that there was a family who created this name and built the guns. The name is now accepted as one of the greatest names in gunmaking, and so I was amazed, one spring morning during 1949, to receive a telephone call at the desk of the shipping firm where I was employed. The caller was my uncle, Victor Seely, who told me that his brother, Hugh Sherwood, had bought Purdey's and would like me to have it. 'What's Purdey's?', I said. 'It is a well known gun shop', was the reply. Strange though it may seem, I had never heard of it. My father had given me a brand new pair of Boss guns for my twenty-first birthday in August 1947, but from my way of thinking a gun was just a gun, made to be used as often as possible. It never occurred to me that there were any great distinctions between one sort of gun and another. 'Uncle Hughie thought it might be rather fun, but there is no need for you to go into it unless you want to.'

I did not enjoy being a shipping clerk, and enjoyed even less the daily life in the Billingsgate area where the offices were situated. The chance of working anywhere else seemed very attractive and so it was arranged that the next morning we would go along to Purdey's in South Audley Street and meet Tom Purdey. I was shown into the Long Room; Tom was charming. He told me about the firm, the families who made the guns, his own family, and introduced me to Harry Lawrence, the factory manager. It all sounded so wonderful that I decided that this was what I wanted to do; and that is how I became associated with the great firm of James Purdey and Sons, gunmakers since 1814, and entered a career so entirely different from any that I had previously planned. It was not until several years later that Harry Lawrence told me that after I had left the building Tom Purdey had said to him, 'Harry, if that little bugger comes to work here, take him over to the factory and don't let him near the shop!'

The circumstances that persuaded Lord Sherwood to buy Purdey's were many and varied. In the 1930s the Purdey family, under strong financial pressure, found it necessary to bring in outside capital in order to keep the firm solvent. A series of financial benefactors and their advisers had

PROLOGUE

been directors, but after the tragic death by enemy action of a new owner, Colonel Ivan Cobbold, in the Guards Chapel in 1944, the firm was once again on the market with Tom Purdey desperate for support. The time could not have been worse for such an appeal. In 1945 the Labour Party came firmly into power, taxation reached over 19s 9d in the £ for the very rich, and all men believed that the day of the big estates, and therefore big shoots, was about to end; very expensive guns would no longer be in demand. It was also believed that the old-fashioned gunmaker was dying out, that young men would never again submit to a long period of apprenticeship, and in any case modern technology would make such traditional skills redundant. Lord Sherwood thought differently, and was proved right.

James Purdey & Sons, Ltd.

Audley House, South Audley Street,
LONDON, W.1.

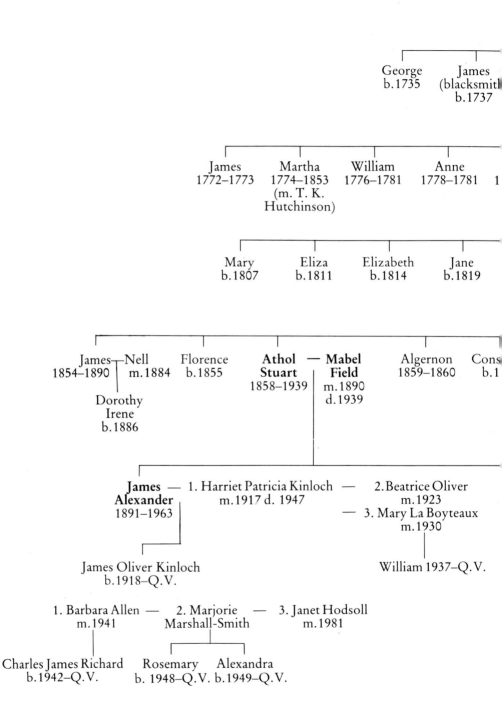

b.c1690

Anne   Robert
        b.1740

33   **James** — Mary   Susanaha
      (The 1790–1860   Feb 1786–
      Founder)   July 1786
      1784–1863

s (The Younger) — 1. Caroline Thomas — 2. Julia Haverson
1828–1909   m.1851 d.1870   m.1873 d.1911

Cecil   Percy
1865–1943   1869–1951

Ada   Talford   Sefton   Mabel   Archibald   Lionel
June 1875–   b.1876   b.1877   b.1880   b.1881   1886–1959
July 1875

ias
ıld
rt
957

ily Tree
ed with the firm are in bold

THE FIRST JAMES PURDEY

According to Purdey family tradition, the ancestor of the gun-making Purdeys, John, walked from east Scotland in 1690 to find work in London. Purdeys, or Purdies, were living in Stepney and Wandsworth in London as early as 1621, and it may be that he was coming south to join relatives who had prospered and held property in that area.

John Purdey kept a record of family matters entitled 'John Purdey: His Book'. This has since disappeared, but a copy of an entry from the book in the Family Bible tells us that he had three sons: George, born 1735; James, born 1737, who worked as a blacksmith in the Minories which was then the gunmaking quarter of the City of London; and Robert, born 1740. Blacksmiths at that time carried out much ancillary work for gunmakers, particularly as barrel-makers.

James Purdey married a girl called Anne. There is no record of her maiden name and in the Family Bible his grandson, James the Younger, writes: 'The above mentioned, James (my grandfather) married Anne Purdey but where not mentioned.' They had three sons and four daughters. Child mortality was very high and the first son, James, born in 1772, died the following year; the second son and two daughters died before reaching the age of four. But James, the third son, and second of that name, born in August 1784 and christened in St Mary's, Whitechapel, survived to found the firm.

It was his elder sister Martha, born in September 1774 and baptised on 30 October the same year at St Botolph's without Aldgate, who was instrumental in starting him on his gunmaking career, for on 13 June 1793 she married Thomas Keck Hutchinson, gunmaker, at St Mary's, Whitechapel.

Thomas had been apprenticed as a gunmaker to Stephen Sandwell in 1785, and become a Freeman of the City in 1792. The marriage was extremely happy and between 1794 and 1817 the couple had nine children, five girls and four boys, eight of whom survived to reach adulthood. In spite of the difficulties involved in bringing up this large brood, Thomas agreed to take Martha's brother as his apprentice, and on 21 August 1798 James Purdey was indentured to his brother-in-law at the age of fourteen.

James Purdey (1737-?), blacksmith in the Minories, and father of the founder

In those days an apprentice lived with his master during his term of apprenticeship, which lasted for seven years, so young James Purdey would have lived with his sister and his nephews and nieces at their house in St Botolph's, Moorfield, which they owned until 1800. They then moved to the Minories (Whitechapel) until 1802, Thomas carrying on business from a gun shop at 232 Borough High Street, Southwark, between 1802 and 1808. They lived at 3 Red Lion Street, Southwark until 1811, then finally moved to Tothill Fields close to the present Victoria Station in 1812. Apart from his gunmaking activities Hutchinson went into the property market, and owned houses in Southwark at Kent Street, Blackman Street and Lamb Alley. Two of the Hutchinson boys, William and Thomas born in 1807 and 1812 respectively, became gunmakers, but were not apprenticed to their father or taken on at any time by their uncle.

During his apprenticeship James would have been taught every aspect of gunmaking and, since guns at that time were flintlocks, he may have been taught, among other things, to forge the 'Damascus' barrels, using nails from old horseshoes heated up and hammered together into strips which were then beaten round rods. (Nails from old horseshoes were supposed to have been toughened by the heavy wear they had to endure.) He would have learnt to file the beautiful 'dogs' and springs, to fit and carve the stocks and possibly a certain amount of engraving. He was bound to serve his master in all things, not to marry or commit fornication, not to gamble or frequent theatres or taverns during his term of apprenticeship, and in return his master fed him, clothed him and gave him lodging and all other necessities of life. Wages did not come into the contract.

James completed his apprenticeship in 1805 and found his first employment with the greatest English gunmaker, Joseph Manton, whose premises at 314-15 Oxford Street was the acknowledged Mecca of all true connoisseurs of sporting guns. While with Manton, he greatly increased his knowledge of gunmaking and at the same time impressed his employer and his employer's customers, for he rose to the position of head stocker. His admiration for Manton's ability was unbounded — 'But for him we should all have been a parcel of Blacksmiths' — and there is no doubt that his contacts with Manton's influential customers laid the foundations for his success when he eventually opened his own business.

In 1808 he transferred his services from Manton to the Reverend Dr Forsyth of Piccadilly. Forsyth, who had been born in 1768 and had become minister of Belhelvie parish, Aberdeenshire, in 1791, was a man of great learning who from an early age was fascinated by firearms and the chemical properties of explosives. He had built a workshop in the grounds of his manse at Belhelvie where, in the course of his experiments — one of which had blown him out of his hut — he noticed that the new powders with which he was working as primers for guns, could be ignited by detonation. He invented a new lock to work on this principle. In 1806 he travelled to London and showed his invention to his cousin, Henry Brougham, the future Lord Chancellor, who in turn introduced him to Sir Joseph Banks, a friend of Lord Moira, then Master General of Ordnance. The latter, realising the military possibilities of this invention which meant that guns could now be fired in heavy rain without the danger of the priming powder becoming wet, thus doing away with the necessity of having to 'keep your powder dry', installed the reverend doctor in the Tower of London and supplied him with workmen and materials. Dr Forsyth continued there with his experiments until 1807 when Lord Moira's successor, the second Lord Chatham, closed down Forsyth's workshop.

In June 1808 Dr Forsyth and James Brougham, brother of Henry, opened a gun shop at 10 Piccadilly to make firearms based on the principles put into practice during the year in the Tower. Joseph Vicars, who had been chief mechanic in the Tower, was in charge, and James Purdey was appointed stocker and lock-filer. Purdey learnt all he could from his employer and furthered his contacts with the trade and his master's customers, becoming a Freeman of the City of London on 31 December 1812. At that time he was living in 15 New Greycoat Place, Tothill Fields.

In 1814, James Purdey opened his own shop in 4 Princes Street, off Leicester Square. He built single and double flintlock guns, duelling pistols and flintlock rifles. He also experimented with the new detonating system, later known as 'percussion', which he had studied while working with Forsyth. He had made a reputation as an excellent craftsman during his employment with Manton and Forsyth, so his name was already known in shooting circles. Colonel Hawker, the great sportsman and acknowledged expert, knew all about him, and in his famous book *Hawker on Shooting* (1844 edition) he quoted Joseph Manton as singling out Purdey as being the best workman in the trade:

> Mr Purdey has still perhaps the first business in London, and no man better deserves it. I once asked Joe Manton whom he considered the best maker in Town (of course excepting himself), and his answer was 'Purdey gets up the best work next to mine!' This was when Purdey occupied a small shop in Princes Street.

Joseph Manton died on 29 June 1835 aged sixty-nine, and Hawker makes no bones about his death marking the end of an era of the greatest craftsmanship in gunmaking: 'The Gunmakers, in short, still remain . . . like the Frogs without a King.' He then goes on to give the most illuminating and amusing description of the personalities and quality of the gunmakers still in business. He states that it is entirely their own fault that their businesses are running down: they have adopted the detonating system which gives a quicker rate of fire than a flintlock and 'nowadays every common fellow in the Market-Town can detonate an old musket and make it shoot as quickly as can be wished', whereas in the building of the flintlock only the goodness of the work could guarantee the quickness of the firing. He summarises the points necessary in a flintlock gun as being:

1 Soundness and perfect safety
2 The barrels being put together for accurate shooting
3 The elevation being mathematically true and raised strictly in proportion to the length of the barrel
4 The stock being properly cast off to the eye, and well fitted to the hand and shoulder

He continues: 'I say nothing of the balance, because any good carpenter, with some lead and a centre-bit, can regulate this to the shooter's fancy.'

The men who had been employed and trained by Manton were all workmen of supreme ability and so, in due course, they started business on their own account — Charles Lancaster, the great barrel-maker who stamped his work with the initials C.L.; Lang of Andover; Moore, who was considered the best stock fitter for Manton; and William Grey, Manton's clerk and manager. Thomas Boss who had learned his trade from his father and then became one of Manton's leading craftsmen, actually worked for Purdey from 1817 until 1821 before starting his own business in 1830 at 3 Grosvenor Street. Incidentally, history repeated itself in 1891 when Boss's nephew took into partnership John Robertson, who had worked for James Purdey the Younger for ten years.

This rivalry among gunmakers is further evidenced by *Pigot's Commercial Directory for Merchants, Manufacturers and Traders in London* (1828–9) which lists the following numbers of firms engaged in the craft:

- 88 gun and pistol makers
- 4 gun barrel makers [including Lancaster]
- 2 gun flint manufacturers
- 5 gun implement makers
- 4 gun lock makers
- 2 gunpowder flask makers
- 9 gunpowder makers
- 5 gun stock makers

And out of all the names involved only that of Purdey is still in independent existence today. So it was in a climate of scepticism by the 'experts' like Hawker, and intense competition by excellent craftsmen bent on making their own way, that James Purdey, had to fight to make his business prosper.

The earliest ledgers and dimensions books listing the transactions and details of the first flintlock guns and pistols are missing from the Purdey records. Mr Tom Purdey, great grandson of the founder, said he had the original ledger in his possession in 1940 — a small black book in which James had kept his own accounts and noted down the customers and details of the guns, together with their numbers, he had built for them. However, thorough search has failed to discover its whereabouts.

Even sadder is the absence of any of the original letters, or letter books, written by the founder, his son and grandson. The firm naturally kept all the ledgers in perfect order, but copies of letters were only retained for twenty years. However, certain special letters written by these three proprietors, were still in existence in 1939, bound in leather and kept in the shelves behind the clerks' desks in the 'front shop'.

Mr Eugene Warner, who joined the firm on crutches in 1918 after being severely wounded as a sergeant of a gun team serving in the Royal Artillery during the 1914–18 war, and who is still a mainstay of the firm to this day, when asked about the whereabouts of these records said: 'Oh, it was all because of that dreadful Lord Beaverbrook.' In November 1939, Lord Beaverbrook appealed to the nation to hand over all old newspapers, letters and spare paper to be recycled for the war effort since pulp could not be brought into England due to the activities of German submarines. At that time, the conviction that everything that could be of help in defeating Germany, however precious, should be gladly surrendered to the war effort was universal. Tom Purdey, having heard this appeal, walked out into the front shop and said, 'No one will want to read all those letters in the future. Put them in sacks and send them off to the paper mills for recycling.' Mr Warner and Graham Tollett, his assistant, did as they were told and so these priceless records were destroyed.

The earliest entry in our Ledgers is for the year 1818, and from then on there is a complete record of the day-to-day transactions of the firm. The names of the customers and the gun numbers of the weapons they ordered are given, together with all financial dealings and discounts; by combining these details with the gun specifications in the dimensions books we can arrive at an accurate picture of the firm's progress. As no letters or workmen's wage books remain before 1863 we have to base the firm's history entirely on these entries. Sadly all entries for customers from 1818 to about 1823 bear references to earlier transactions in the lost book and give page numbers and letters accordingly. All are in lovely copper-plate writing.

The first entry covers the sale of a double-barrelled detonating or percussion gun on 18 April 1818 to Lionel Goldsmid Esq, for £52 10s. On 7 December 1819 the same gentleman bought a pair of extra barrels for his gun for £16 16s and on 4 January 1820 a single gun in case for 18 guineas. Various other transactions continued until 1822 when his account was closed for the time being. Then in 1849 he ordered a best double gun, No 1709, for £55, complete with case and fittings. Alongside in Mr Purdey's writing are the words 'The Bills dishonoured. Bankrupt' and a note 'By Cash from Insolvency Court by C. Wick Bernordy: £20 3s 10d'.

The next entry is for December 1818, when a Major Wellington took delivery of two extra springs and hammers for £2. Presumably the major failed to settle his bill as his entry is crossed out twice and James has written 'BAD' in big black letters alongside the amount. This treatment is also accorded to the Earl of Belfast in May 1822 for a sum of 7s 6d for the shortening of his stock, and Earl Fortescue gets a 'doubtful' in 1829 for 17 guineas. James Purdey eventually employed a solicitor, Mr W. A. Goatley of Cork Street, to collect his outstanding debts. There are separate pages covering his activities at the back of each Ledger.

THE FIRST JAMES PURDEY

The first entry in the Dimensions Books, dated 1820, shows percussion gun No 402. Where complete, details give the gun numbers, strictly in numerical sequence, together with the type of gun and number of barrels, either single or double, the bore and the initials of the craftsmen who built the weapon at each stage of completion. The stages include jointing; locks jointed; stocking; screwing together; marking of crests and escutcheons; cocks fitted; detonating and finishing; polishing; engraving; regulating and finishing; varnishing, browning and casing. After this comes the name of the customer. It is thus possible to check on the builder of any part of a gun, and also to make an accurate count of the number of guns, rifles and pistols sold in each year. Until 1828, the entries were not kept up to date, so it is not always possible to find full information on guns, rifles or pistols built before that time, but a Mr C. B. Calmody took delivery of a best finished rifle No 1406 in 1828 and from then on most details are faithfully recorded.

Purdey guns were produced in every conceivable bore size: 10, 11, 12, 13, 14, 15, 16, 17, 17½, 18, 19, 20 and 22. The 17½ bore must have been James' own special size as there is no record at the London Proof House of such a gun being proved or stamped. (The criterion for measuring bore size was based on the number of lead balls which fitted the internal diameter exactly and made up the weight of 1lb.) Stock lengths were between 14in and 14½in from the front trigger to the middle of the stock; but a Mr Miller, who must have been a tall man for the time, took a 14⅞in stock for his 14-bore gun in 1823.

It was the practice to convert flintlock guns to percussion, but I can find few references to this having been carried out in Purdey's workshop, although on 21 October 1823 the Reverend J. Cheshire, vicar of Bennington, Stevenage, had his single gun 'altered to the detonating principle at a cost of £3. 3. 0, and a new mainspring for £1. 3. 6'.

Apart from building new weapons, James Purdey maintained his customers' guns, sometimes on a yearly basis, carried out repairs and supplied new parts, and also sold shooting equipment and accessories. Some early charges are:

1820

Cleaning and Re-browning Barrels	15s 0d
Cleaning only	15s 0d
New Tumblers, Bridles, Pegs	
1500 Copper Caps	£2 12s 6d
New Ram-rod	5s 6d
A Powder Flask	9s 6d
Regulating Shooting & Boring Barrels	15s 0d
1 lb of Wadding	11s 1d
3 doz. Cartridges (shot wrapped in wire netting)	6s 0d

Cost of new weapons was:

A new Best single Gun	£52 10s 0d
A new Best double Gun	£55 0s 0d
A new Best small Rifle (single)	£36 15s 0d
A new Best double Rifle	£70 0s 0d
A pair of double Pistols. Cased	£26 15s 0d

From 1823 onwards, more and more sportsmen were ordering percussion guns. Lords Belgrave, Barnard, Lilford, Gosford and Belhaven all ordered guns and rifles in that year, and Lord Blandford bought the first of many guns and rifles that the Churchill family would acquire when he ordered double-barrelled 20-bore percussion gun No 639 with 30in barrels.

Lord Middleton, who had ordered a pair of 13-bore single-barrelled guns Nos 744-5 in 1824 for £52 10s, acquired a 14-bore single-barrelled rifle No 827 with 29in barrel in 1825 as well as a double 16-bore gun No 862 and a double 13-bore No 911. In 1826 he ordered two more rifles of unknown size Nos 1006-7, and in 1827 a further gun No 1218 for £55. In all he took delivery of 22 guns, 6 rifles and 2 pistols, one with four barrels, between 1824 and 1829. Lord Middleton had a large armoury at Wollaton Hall, his home in Nottinghamshire, including a small cannon which was mounted on the roof and used by the gamekeepers to ward off attacks by Chartist rioters from Nottingham. Lord Hopetoun was also a great supporter, ordering 4 guns, 2 rifles and 3 pistols during the same period.

Customers increased yearly and included a satisfying number of clergymen. Fifty different reverends were regular customers between 1822 and 1827, buying new guns and maintaining old ones, including the Dean of Exeter and the Reverend Edward Missenden Lane of Somerly, Yarmouth, who on 17 November 1827 brought in his double flintlock gun No 224 to be cleaned for 5s, the barrels repaired and re-browned for 15s, and was supplied with 50 flints for 2s 6d.

The first member of a royal house to patronise James Purdey was His Serene Highness the Prince of Leinegen, who in 1823 ordered double-barrelled 14-bore percussion gun No 817 with 30in barrels. Shortly afterwards, in 1825, the Duke of Gloucester, brother to King George IV, ordered a double-barrelled percussion 16-bore gun with 31in barrels No 856, and later in the same year took delivery of a 15-bore gun No 945. The cost of each of these guns, including case, was £55. Both guns were completely overhauled in 1828, and thereafter on regular occasions. The Duke was also the owner of a Manton 'copper tube' gun for which Purdey's supplied the tubes. The Duke of Cumberland, his younger brother, bought his first pair of guns in 1829, and was followed by the Crown Prince of Bavaria and the Prince of Orange in 1836.

Apart from building new guns, James also dealt in secondhand ones, and in guns of other makers. Edward Bray bought a secondhand single-barrelled gun in 1825 for £25, and Sir John Conroy one for £18 13s. Guns of foreign build were also maintained and repaired, and guns of all makes were stored on the premises. Part exchanges were frequent: H. Wodehouse Esq bought gun No 1755 for £30 in 1826 at the same time trading in his old one. Guns were also built for the trade, two of the main recipients being Joseph Lang of 7 Haymarket and Charles Lancaster who opened shop in 1826 at 151 New Bond Street. Purdey also sold Lang flintlock pistols, and guns by Forsyth, Joe Manton, Egg, Wilkinson and Wigram, as well as guns of his own early make.

He started to build up his trade in India in 1826. The new percussion mechanism must have been popular there and, having opened negotiations with Messrs Middletons of Calcutta (Allardice), he sent out 4 best double guns at £63 each with 20,000 copper caps in 1826, and followed this up, two years later, by shipping 12 guns and rifles with 30,000 caps. This was to lead to a growing demand from Indian Army officers and, later, Indian princes.

Although he built beautiful guns and pistols, James's main love was rifles and he excelled not only in building them but also, as a marksman, in testing their shooting. Among other sizes, he built 16-bores, shooting 1½ drams of powder with a 1oz round ball with deep rifling of 10 grooves, and a 32-bore using 2¼-2½ drams and a short conical bullet. His 2-grooved rifle, which fired a conical bullet with two step wings, gave high velocity and low trajectory at good ranges. He tested the accuracy by shooting at small plaster casts shaped like 'butterflies'.

By 1826, with a great reputation in the sporting world, he must have felt financially secure, for on 1 August he acquired the shop and workshop premises of his old master, Joseph Manton, at 314½ Oxford Street, now covered by D. H. Evans, the famous department store. The rent was £260 per annum and the premises were insured for £1,200. This was the most famous gun shop in London. The year 1828 brought great personal happiness. We do not know the date of his marriage as his son, James the Younger, wrote in the Family Bible, referring to his father: 'The above mentioned James married Mary Purdey, but where not mentioned.' Nor do we know her maiden name. They had five daughters, the first, Mary, being born in May 1807 at '01 minutes past one in the morning' and christened in Whitechapel. The next three daughters were born in 1811, 1814 and 1819, the exact minute of birth being faithfully recorded in each case, and were christened in St James's, Westminster. Finally, at '25 minutes past two o'clock' in the morning of 19 March 1828 the longed for boy, James, was born and on 3 August was christened at St George's, Hanover Square. So having assured the continuity of the family and secured a firm base for the gunmaking business, James Purdey was well set for the future.

With more space available in the new premises, he expanded his trade. Apart from gunmaking, and the provision of shooting accessories of all types, he continued to build guns of lesser quality which he sold from his shop or supplied, at a low price, to members of the trade. His best small guns at that time cost £57 15s each, a best finished large double gun £63. John Baker of 6 John Street, Bedford Row, was sent 15 new guns of various sizes in 1829 at a cost of £40 per gun in its case, and in 1831 received another 6 guns on the same basis. James Collis, gunmaker, of 12 Vigo Lane off Regent Street, dealt in 31 new guns, rifles and pistols on a sale or return agreement between 1830 and 1835 at £32 per gun.

Between 1830 and 1834 Joseph Lang was supplied with eighty-four guns, rifles and pistols, the gun numbers lying between 1760 and 2560. The guns at £32 each, sale or return, were with one exception all sold within three months of delivery as against each entry is the date on which the guns were paid for. One gun, however, did not sell for five months and on 4 January 1832, Purdey finally took settlement in kind, namely 'By a bay mare £32'. In fact he was not at all particular as to how he was paid, provided he got his market-price equivalent. In 1828 a Mr Hitchcock could not settle his bill in its entirety and James agreed the following deal:

12 chaldon corals	£32 19s 0d
By a gun	£21 0s 0d
By allowance	£6 4s 0d
By wine	£14 10s 0d
By coals (as at August 31st 1826)	£39 6s 6d

and finished off the arrangement by selling Mr Hitchcock a leather cover, value £1, in exchange for four bottles of wine.

Mr Avery, New Bond Street, owed £395 6s in 1833. Eventually Purdey had to accept a loss of £98 10s but he took in kind a gold chain, two pairs of candlesticks, a snuffer and tray, a silver teapot worth £8 10s, a silver cream jug, a silver sugar basin, and a cash dividend of £21 10s. The entry ends: 'Signed an agreement relinquishing all further claims on receipt of the above dividend. Oct. 9th. 1833. by Mr. Purdey.' In 1832 another horse was acquired for £136 10s from A. McKenzie Esq, Regent's Park.

'Gamekeepers' guns' were built and sold regularly, the price in 1834 for a gun of this quality being £20, spare mainsprings and cocks £3. Lord Ducie of Woodchester Park, Gloucestershire, ordered the first gun of this sort for his keeper, Mr Blackburn, and the consignment cheque amounted to £3 5s. A Mr Kenyon did likewise in the same year, and several such guns of 14- and 13-bore calibre were supplied to Charles Lancaster.

Rifle clubs were starting to be fashionable and Lord Kennedy, captain of the Aberdeen Rifle Club, took delivery on 16 February 1829 of a large rifle in case, No 1199, for £36 15s, with a sling 10s 6d, balls and patches 15s, patch case 10s 6d, and 500 caps 17s 6d. In the same year James Purdey gave Lord Henry Bentinck a double gun, No 1265 — a 14-bore with 30in barrels. This gift may have a bearing on the kindness Lord Henry rendered to the Purdey family, eighteen years later, when the latter were in financial difficulties.

Several customers appear to have been in the habit of paying their bills on credit or, in modern terms, on the 'never never'. In 1828 Sir H. Calder was entered as 'Will pay in July 1833; Will pay in Oct. 1833; Will pay in Dec. 1834'. The total amount involved was £68 10s and, with this long time for payment, cash flow must have become strained.

Nevertheless, Purdey continued to make guns of such quality that orders poured into his shop from men like William Beckford of Fonthill, Captain St Leger, Lord Ilchester and Sir James Boswell of 44 Queen Street, Edinburgh, who in 1826 bought a double gun for £57 15s. The Earl of Airlie bought a 17-bore rifle on 10 October 1829 for £39 18s, and was soon followed by his Ogilvy brothers in their purchases of guns, rifles and pistols. But the sale of guns and shooting accessories was not the only source of trade; James dealt in much besides. The 1828 entry concerning the Reverend George Watson of Earlstone Park, Devizes, and 6 Grosvenor Square, includes 'Paid carriage for a Grouse 4s 6d.' Also in 1828, Major Paul of Calcutta was supplied with 'A pair of Best Razors 17s 0d, and a Double bladed Penknife 5s 6d'. And in 1830, to the same officer: '6 dozen Salmon Hooks and Floats and a Coil of Salmon Gut, £1 18s 0d.' In 1831 a gun was lent to Lord Cranston for the season for £5 and in 1836 the same customer was allowed 5 per cent on an outstanding bill of £286 14s for four years — perhaps one more explanation for the financial difficulties to be experienced by James in 1847.

Military type weapons were also supplied to the trade and the army. In July 1831 Messrs Thompson and Poole, of Broad Street Buildings, ordered a carbine and bayonet case (No 2149) for £29 5s 1d, plus two daggers at 63s each, and a pair of pistols with 7in barrels (Nos 2147-8) for £42. In the same month the Reverend J. Pettat of Southtrop House, Fairford, Gloucestershire, had his flintlock gun (No 357) done up, the amount being settled by the 7th Hussars. William Gambier Esq of Sacombe Park, ordered 6 fuzees with bayonets at £15 and 12 cutlasses at £3; and Lord Tankerville ordered 30 musket rifles (Nos 1-30) with bayonets at £10 5s each. The Reform Bill controversy was then at its height, and these gentlemen may have been laying in arms to protect their property from the reforming bands which were roaming the countryside. If so, it is satisfying to find that Sir Robert Peel, Bart, Whitehall, became a customer in 1832.

Purdey supplied Lord Castlereagh with 6 guineas' worth of the best cigars the same year, and sold several cast-iron rabbit targets at £1 10s each.

He kept a dog for J. Sherwell for nine weeks at 13s 6d, and bought a dog for the Honourable G. Cadogan in 1836 for 3 guineas. He supplied Mr Bishop of Westley Richards, New Bond Street, and William Moore and John Manton with cartridges in 1832–3. He mended a fishing rod for Mr Higginson and sold him a blue-light flare for 1 guinea in 1832.

The increasing number of fatalities and woundings due to malfunctions in guns, or careless handling, had been highlighted in 1826 when the Reverend John Somerville, minister of Currie in Midlothian, who had invented a safety slide which locked the triggers of a gun when the slide was drawn back by the shooter, published a book called *Essay on the Safety Gun, by the Inventor* dedicated to the Duke of Buccleuch. In this he outlined his invention, then published letters of appreciation from several eminent Scottish gentlemen and summarised the details of accidents caused in the field or in the home. Of the one hundred accidents listed, no less than twenty-one concerned children under the age of seventeen. The dangers of leaving loaded guns near children is instanced by the case of a boy of twelve at Sittingbourne in Kent, who picked up a gun which had been left loaded, and playfully pointed it at his friends. After several ineffectual attempts to pull the trigger the gun went off and killed two of them on the spot. The danger of crossing a hedge with a loaded gun is cited in the story of Lord Albemarle's two sons, Charles and Edward, both under the age of eighteen. While shooting at Bungay in Suffolk on 27 September 1817, Edward's coat caught the trigger and the contents of the barrel lodged in his brother's leg, who bled to death before help could be found. Edward lived on to the age of eighty-three, having taken Holy Orders as a result of this accident, and died as Rector of Quidenham, the Albemarle living in Suffolk.

As a result, James Purdey started to fit safety mechanisms. In 1827 an entry reads 'Two new safeguards to double guns, Nos 1041-2, and the locks of these guns were altered to the safety principal for £11 0s 0d, for W. R. Jones, Esq'; in 1831 he fitted his new patent safety guard and safety stop to a 16-bore gun for Sir Charles Mills. On 13 September 1831 he fitted the new guard and safety stops onto gun No 1967 owned by John Barham Esq for £8 8s. Mr Barham settled his total account by selling gun No 1975 for £26 8s and making up the difference on 18 November 1831 'by a horse for £136 10s 0d'. Mr Purdey's stables were filling up!

The sport of pigeon shooting was gaining popularity and, apart from supplying live pigeons as targets, James also supplied live sparrows, as is shown by the entry for 30 August 1831 which states that sparrows for L. Stephens, Esq of 32 Portman Square were supplied for use at the shooting grounds at an unspecified sum.

Red glass balls, the forerunners of clay pigeons, cost £1 16s for 3,000, and were supplied to a gunmaker, Spencer of Northallerton, in July 1831.

Spencer's went bankrupt in 1837, and cost Purdey's £2 10s 6d. These glass balls were thrown by a machine for the sportsman to practise his aim, on the same principle as the present clay-pigeon trap. Shooting-ground charges in 1830 amounted to 6 dozen shots for 12s, 3 dozen for 6s; and live pigeons were supplied at £4 for 4 dozen birds between 1831 and 1837. A pigeon trap cost £1 10s, and a basket for carrying pigeons 14s.

But one of the most exciting entries is dated 6 September 1831. The significance of this sale cannot have been obvious to James Purdey at the time, but its outcome was to revolutionise the thinking of the whole world. On that day Captain Robert Fitzroy, RN of HMS *Beagle*, took delivery of the following:

A double rifle 2107. 18-bore 28in barrel	£73 10s 0d
A double gun 2136 14-bore 30in barrel	£57 15s 0d
Double pistols with back actions 2141-2 (no bore size)	£47 5s 0d
Pair of pocket pistols (no number, no size)	£8 8s 0d
Pair of duelling pistols (both numbered 1706) with holsters and leather covers	£60 0s 0d

Alan Moorehead in his excellent book, *Darwin and the Beagle*, gives a description of Captain Fitzroy taking Charles Darwin up to London to buy the guns and supplies necessary for this historic voyage. As they would be away for several years, spares were provided for all eventualities and the list continues as follows:

2 spare mainsprings each for all guns and pistols at 15s each
Sear springs 15s each
Cocks £3 each
Pegs £2
20,000 caps £35
24lb of wadding £12
100lb of powder 12 guineas
24 dozen cartridges £2 8s
6cwt of shot 25s
Assorted flasks, covers etc.

The total bill came to £377 9s. Purdey received £215 in cash and gave a discount of £12 9s. The remainder was to be paid by a bill of £150 at twelve months' notice. After making these arrangements, Captain Fitzroy and Charles Darwin visited the ship at Plymouth on 11 September and eventually sailed on 27 December. They were not to return to England until they landed at Falmouth on 2 October 1836.

The payment of debts was, however, one of pressing necessity. We have

Dr Captain Robert Fitzroy R.N. H.M.S. Beagle

				£	s	d
1831 Sept	6	A Best Dble Rifle mahogany case complete	2107	73	10	
		Leather Cover		2	5	
		A Best Dble gun mahogany case complete	2136	57	15	
		Leather Cover		2	5	
		Case of Dble Pistols	[illegible]	47	5	
		Pr Pocket Pistols		5	5	
		Leather Cover		1	16	
		Duelling Pistols in case complete	1708	30		
		Leather Cover		1	1	
		Duelling Pistols in case complete	1706	30		
		Leather Cover		1	1	
		3 Live Leather Covers		4	14	6
		Pr Pistol Holsters		2	2	
		2 Cocks		3		
		Pr Main Springs			15	
		Pr Springs Sear	Dble Rifle		15	
		10 Pegs		2		
		2 Cocks		3		
		Pr Main Springs			15	
		Pr Sear Springs	Dble gun		15	
		10 Pegs		2		
		1 Cocks		1	10	
		1 Main Spring			7	6
		1 Sear Spring	S Rifle		7	6
		5 Pegs		1		
		2 Pr Cocks		6		
		2 Pr Main Springs		1	10	
		2 Pr Sear Springs	Pistols	1	10	
		10 Pegs		2		
		2 Cocks		3		
		Pr Main Springs	Duelg Pistols		15	
		Pr Sear Springs			15	
		20000 Caps		35		
		24 lb Wadding		12		
		18 lb Powder		12	12	
		carried forward		353	7	6

Dr Captain Robert Fitzroy R.N. Cr

			£	s	d				£	s	d
1831		To Amount brot forward	353	7	6	1831					
Sept	5	6 Bags Shot @ 25/	7	10		Sept	5	By Cash	215		
		2 Doz Cartridges	2	8				Discount	12	9	
		Bag [illegible]		10	6			Bill at 12 months	150		
		6 Brass Chargers		15							
		2 [illegible]	1	12							
		6 Pocket Powder [illegible]	4	16							
		2 Canvas Flasks	1	12							
		Bottle of [illegible]		16							
		Browning	2	12							
		Packing Case	1	1							
		2 Spare Rammers		15							
			377	4	6				377	4	6

The Honble [illegible] Rice Trevor

				£	s	d			£	s	d
1831							1832				
Sept	6	A Best [illegible] Rifle [illegible] case complete	2[illegible]	39	18		[illegible]	By Cash	42	3	
		Leather Cover		2	5						
				42	3				42	3	

seen how James had been forced to employ a solicitor to help him in this respect, but the ledgers contain copies of several letters both to and from defaulters. T. B. Arundel, Esq of Sandy Park, Chagford, Exeter, owed £147 3s 2d between 1836 and 1839. Purdey wrote to his 'man of business', Mr Maxwell of north Devon: 'Sir, I beg leave to acknowledge receipt of your letter and would accede to these terms you propose, as I do not wish, under the circumstances, to be harsh with Mr. Arundel, but I hope the Installments will not run over any lengthened period.' The account was not finally settled until 1846.

Captain Virginius Murray had been ordered to the West Indies at three days' notice in 1846, owing Purdey's £76. On 18 February 1847 he had the courtesy to write: 'I hope, Mr. Purdey, that you will believe how deeply I regret having so long been in your debt. The fact is, years ago, I contracted debts from which I am not free, yet my debt to you has caused me much sorrow for I fully appreciate the considerate manner in which you have behaved towards me.'

An equally honest customer who, in spite of being in financial straits did what he could to recompense his gunmaker, was Sir George Leeds of Croxton Park, St Neots, his family's home since 1570. Having been an excellent customer since 1827, he fell on bad times in 1830, and had to sell his lands and house. Purdey's entry reads 'BAD' and 'Heard that he is abroad. Loss £37 16s 4d.' However, Sir George returned his guns to Purdey who recouped his losses by selling them to other customers. A pathetic letter of appeal from a Mr John Anderson of Devon over the payment for a new gun, tells how he is quite destitute and being supported by his daughter: 'Mr. Purdey, I solemnly assure you, as I hope for Salvation, that I am in that Distress that my situation is most deplorable, since at the present time I do not receive a shilling neither have I done so for a long time past.' This was as a result of his properties in the West Indies running into debt. Purdey has written on the bottom of the letter: 'A Bad Job.'

Mr Goatley had twenty-eight bad debts between March 1840 and July 1841, and seventeen in 1844, a difficult one being that of Captain Skinner from Poona who declined to accept two rifles because of delays in Customs formalities. There was frequent trouble with the authorities in India, and after the Honourable P. J. Pellew of the 7th Madras Light Cavalry had had three pairs of double pistols and ammunition sent out to him between 1842 and 1845, the entire consignment was returned in 1847 costing Purdey £180 9s. Captain Scott Thompson of the 7th Dragoon Guards who had owed £30 17s 6d since March 1848 had a nasty shock: 'Threatened him by Letter, Feby. 11th 1856 with Law. He is now a Colonel at the Horse Guards.' An

The entry in the firm's Ledgers for Charles Darwin and Captain Robert Fitzroy of HMS *Beagle*, 6 September 1831 interest rate of 5½ per cent was charged on all outstanding accounts, and an agreed rate for credit. Lord Villiers agreed to 5 per cent on a bill of £268 in 1845 after he had 'Promised to Pay Mr. Purdey'.

On the other hand, mistakes in the accounts department did occur, and opposite a bill for £1 10s 6d in 1836 for Captain Wemyss of Wemyss Castle, Fife, James has written: 'Must be somebody else. The above Gentleman does not acknowledge the Debt.'

In the meantime the ordinary day to day dealing continued. In 1834, Lord George Harvey returned his brand new gun to be rebuilt, as it was not good enough; and I am afraid modern perfectionists will be appalled to hear that, in 1836, James Purdey was selling furniture polish for doing up his stocks at £1 18s 1d per bottle. The stocks were made of walnut, sometimes 'ebonised' with a black lacquer finish, but mainly finished with oils. Bird's-eye maple was also used and gave a lovely black and gold look to the gun.

Lord Cardigan, of Balaclava fame, bought his first 14-bore rifle for £84, and Lord Vernon got Purdey's to 'do up' his wonderful collection of guns, pistols and rifles of every conceivable English and foreign make. Between August 1836 and February 1837 twelve pages of the ledgers are devoted to this one operation alone with over 360 specific entries in the year; the total cost amounts to £723 — a huge amount for that period. The pistol trade expanded; in 1831 Lord Hopetoun added to his collection by ordering a four-barrelled pistol in case for £45, and Lord Middleton went one up by taking three four-barrelled pistols at the same price. The Duke of Newcastle not only had a pair of similar weapons, in case, for £94 10s, but also a pair of pocket pistols. Iron targets for practice with these weapons were available at £2 10s, presumably in the shape of a man. The first revolving pistol was sold in November 1853, for £8 10s plus ammunition, and a saloon pistol for £5. (A saloon pistol was used for target practice in a pistol range.)

But the greatest honour came to James Purdey on 23 October 1838, when Queen Victoria, through Mr Backhouse of the Foreign Office, ordered a pair of percussion double-barrelled pistols for presentation to the Imam of Muscat. The rosewood case was inlaid with buhl and lined with crimson silk velvet, arms were engraved on the centre of the case, with a crown and cypher on the stocks. The numbers were 3151-2, the barrel length 8½in and the price £73 10s. In her Journal of 20 September, the Queen recorded: 'Spoke to Lord Melbourne [the Prime Minister] . . . of presents for the Imam of Muscat: I'm to send him my picture set in Diamonds in a Box, and besides a Sword and a Brace of Pistols, which Lord Palmerston [the Foreign Secretary] will get'. The Imam's envoy was in London at the time and brought the Queen a number of gifts, including six horses, to mark her Coronation on 28 June 1838.

On 22 November of the same year, Her Majesty bought a highly finished double gun, gold mounted, No 3039, for £67 10s, and a gold mounted rifle, No 3027, in a rosewood case lined with velvet for £50. In 1839 she purchased a pair of best double flintlock pistols in case for £63. Also in 1839, Count D'Orsay took away a gun, No 1847, for which he did not pay, costing Purdey £57 19s 6d, but both the 'Roi de Wurtenberg' and Prince Jerome Napoleon Bonaparte not only ordered guns, but paid for them.

On 11 July, 1840 Prince Albert ordered a best double gun, No 3309 — a 16-bore with 30in barrels, the latter richly inlaid with gold, the case lined with blue silk velvet — for £68 5s; and a small rifle, No 3259, for £36 15s. The Prince of Orange ordered an exactly similar rifle, No 3310, in September of the same year.

In 1841 James Purdey was elected to the position of Master of the Worshipful Company of Gunmakers of the City of London. This position is held for one year, although re-election is possible, after the candidate has filled the two junior appointments of Upper Warden and Renter Warden. Before that he must have been elected to the Livery of the Company, and have also served on the Court of Assistants, which Purdey joined in 1838.

The Gunmakers' Company was founded by a Royal Charter granted by King Charles I in 1637. Before that time blacksmiths and metalworkers had all tried their hands at making and selling firearms with the result that several people had been blown to pieces or maimed when the weapons had exploded. The reputable gunmakers started to be identified with such practices and consequently got a bad name. The King was therefore petitioned to create a company whose purpose would be to test all weapons to safety limits laid down by an Act of Parliament before such weapons were sold to the public, and the company would have the authority to stamp with its arms and insignia the weapons which passed this proof test. Anyone selling a weapon which does not bear those marks is still liable to prosecution.

The Company consists of a Master, Upper Warden, Renter Warden, and a Court of about sixteen Assistants made up of responsible gunmakers who are elected for their integrity and known desire to maintain the Standards of Proof for the safety of the public. A Working Committee is appointed by the Court from among its members to oversee the proofing of weapons and this small committee is known as the Proof House Committee. From time to time the Rules of Proof, which govern the pressures to be used in the proofing of guns, are brought up to date by Act of Parliament, but guns are still proofed at the Gunmakers' Company in the Commercial Road, on the same site as originally granted. The Proof House was placed thus outside the City walls because of the great risk of fire where large quantities of explosives were concerned. The Court Room is situated above the Proof House so its meetings are disturbed by the crash and bang of exploding proof charges.

It was about this time, even though the business continued to prosper according to entries in the Ledgers, that something went wrong with the finances of Purdey's. The story passed down by the family and told to the author by Tom Purdey, the founder's great grandson, is that the crisis was due to cash flow failure and that Purdey very nearly went out of business. This may have been largely due to the reluctance of customers to pay what they owed, or to James's extravagance in living above his income, but in any case Purdey was saved, so we are told, by the intervention and kindness of Lord Henry Bentinck, a younger brother of the 4th Duke of Portland.

In 1847, Lord Henry Bentinck, walking down Oxford Street, saw Purdey putting up his shutters at the end of the day's work, and asked him how his business was going. James admitted that things were very bad, not only were his customers not bothering to pay their bills, but orders were not coming in as anticipated. Lord Henry is reputed to have said 'leave it to me', and from that time the records show that 125 guns of all kinds were produced each year until 1852. In the Long Room in South Audley Street there is a photograph of Lord Henry bearing a note in James Purdey's handwriting: 'Lord Henry Bentinck. My Earliest and Best Customer and Friend when I was a Youth.' On the back of the photograph, which was taken by Messrs Bassano in their studios at 25 Old Bond Street, there is a further note in the handwriting of the son of the founder: 'The Best and Truest Friend and Patron to James Purdey the Elder and Younger.' No one knows exactly what it was that Lord Henry did, but it must have been very effective, as after this the financial position became much easier.

The main cause of the crisis was probably the long-term credit given to customers who simply did not bother to meet their debts, and these bad debts continued to recur in some measure. Lord Malmesbury owed £1,057 from 1846 to 1848, and Lord Clinton's one debt continued from 1846 to 1851. Lord Sussex Lennox, a director of railways in England, Wales, Ireland and Spain, who was eventually arrested for debt in July 1857 after his companies had gone defunct, came to an agreement on 11 May 1847. 'I hereby agree to pay Mr. Purdey Ten Pounds per annum in Liquidation of my Debt to him of Forty Pounds, the first payment to be made on 25th September for which I enclose a cheque on Messrs. Drummonds. Signed Sussex Lennox.' James Purdey was sent a notice of his appearance in court in 1857, so presumably his debt had still not been cleared.

There is, however, one other possible contributory cause to the crisis — the political and social climate of the times. This was the period leading up to the 'Year of Revolutions' of 1848, and the grave unrest in all parts of Europe and England would have made customers chary of paying large sums on expensive guns. John Arlott, in his history *The House of Krug*, states that the same crisis took place in 1847–8 in the champagne business;

profits fell dramatically and in 1848 ceased to exist, although the following year a more stable climate put things back to normal.

In spite of this setback, however, Purdey commissioned his portrait. This fine picture, attributed to Sir William Beechey, RA, now hangs in the Long Room at Audley House. It shows him looking extremely prosperous in a smart coat with gold buttons and extraordinarily like all his descendants. Perhaps he was the origin of their charm but also their fiery temper, which has so often terrified their employees and even daunted their customers.

Now that trade was starting to expand once more, new contacts were made. The first Russian customer, the Grand Duke Constantine, bought a gun in 1847 and was followed by the Emperor of Russia, Tzar Nicholas I, who ordered four best double guns in June 1850. The Prince Consort ordered four more double rifles in 1848, and Queen Victoria, in July 1849, ordered a pair of panniers with saddle harness, a salt bottle with silver top, and straps etc for £13 4s. The Nepalese Ambassador, Jung Bahador, took twenty-six rifles and pistols in 1849, together with an unspecified number of military rifles and bayonets, and a six-barrel revolving pistol (Adams Patent) in 1853.

It was at this time that an invention completely revolutionised the building and use of sporting guns, altering them to the form in which we know them today. In 1834 a Frenchman, Monsieur Le Faucheux, had patented a gun which opened between the breech and the barrels and led the way to the 'drop-down' or breech-loading gun. Naturally a propellent had to be devised for use in such a different kind of weapon, and a friend of his, Monsieur Haullier, took out a patent in 1846 for a 'pin-fire' cartridge for use in guns of this type — the 'pin-fire' gun fired by a hammer which struck the pin attached to the cap of the cartridge and caused an explosion by friction. These guns and cartridges became very popular on the continent, but in England grave doubts were held as to their safety and old prejudices maintained that the percussion and even flintlock guns were safer, as the barrels and 'action' were fixed permanently together.

However, Le Faucheux and Haullier displayed their wares at the Great Exhibition of 1851 and the Prince Consort showed interest in the invention. Even more important, that great gunmaker Charles Lancaster, having seen and examined Le Faucheux's guns, immediately saw its potential and took out an English patent in 1852 for a 'drop-down' gun. Other gunmakers immediately started to produce their own version of this system and took out their own patents, more one suspects for advertising purposes than with any real hope of being able legally to protect them.

Cartridges were modified and improved, one of the major difficulties being to devise a method of turning over the paper end so that the pellets did not escape in the sportsman's pocket, or were not jarred out of the cartridge case in a double-barrelled gun when the first cartridge was fired. James Purdey the Younger invented a Cartridge Turnover Machine to achieve this end and drew out a patent for his invention in 1861. He sold this machine to his customers, and the rest of the trade, for 12s 6d a time.

In view of his customers' reservations on grounds of safety to this new development, Purdey was cautious before committing himself to building guns of this type, and the first breech-loading Purdey gun was not built until May 1858 when a pin-fire, No 5462, was delivered to J. Cooper Esq. It was a 12-bore, built with 31in barrels, and cost £57 15s; 500 loaded cartridges for £4 10s were included in the order. Mr Soames, Lord Alan Churchill and Mr Andrew Jardine had similar guns delivered in that year, and Mr Davenport Bromley ordered a pair of 40-bore rifles built on the same principle. However, very few pin-fires were made by Purdeys, and where the records state that percussion guns have been altered to pin-fire there is no corresponding account in the ledgers, which shows that these conversions were carried out elsewhere, and by other gunmakers.

A drawback when using pin-fire cartridges was the danger of the pin being squeezed while the cartridge was still in the shooter's pocket and so slowly, from 1861, centre-pin cartridges came to be used and guns had to be adapted accordingly. Various gunmakers, as already mentioned, produced their own systems and designs, but one of the most important points as regards both safety and operation was that the barrels should be held tight against the face of the action when the gun was closed, and that the mechanism which held them in that position be absolutely secure. It was in this field that Purdey was to establish his name in 1863 when he invented and patented the Purdey Bolt, which fulfilled these requirements.

James Purdey had previously tried various methods of closing 'drop-down' guns by levers situated under the action, and thumb levers let into the trigger guards. Sometimes these mechanisms included a safety stop on the triggers, to prevent them being pulled before the gun was completely closed. Eventually his combination of bolt and top lever, patented in 1870, proved to be the simple and superior mechanism that gunmakers had been searching for.

The Rothschild family, Alphonse, Solomon and Gustave, bought guns in 1853; and Sir Edwin Landseer, the animal artist, was lent a double rifle on 19 August 1854, at a rent for the season of £4 7s; the exotic sounding Emperor of Solo, a district in Java, had two best guns in 1856. Prince Duleep Singh and Lord Huntingfield of Heveningham in Suffolk, who later in the century were acclaimed with King George V, Lord Walsingham and Lord Ripon as the five greatest game shots in the world, bought their first percussion rifles in 1857. In the same year the Prince of Wales brought in a German rifle to be cleaned, and granted Purdey his Royal Warrant of Appointment as gunmaker.

At this pinnacle of his career, James decided it was about time he eased off and delegated more responsibility to his son, James the Younger. He was now seventy-three years old and, by his work and skill, from the humblest of beginnings, had made himself the leading gunmaker in London and a household name among the most important shooting circles in England and Europe. In 1857 he acquired the leasehold of a 6-acre plot of land in Harrow called Foxholes, which was situated just behind the Orange Tree beer shop off the Harrow Road. Before this the firm had used fields at Hornsey Wood House for shooting grounds, and continued to do so thereafter for shotgun testing, but Foxholes, leased for twenty-one years, was to be used as a shooting ground and a site for testing rifles, the majority of which were being built on the premises in Oxford Street. Having completed this purchase it was agreed in an indenture in October 1857 between father and son, that James Purdey the Younger should take over the full running and ownership of the business as from 1 January 1858.

'Young' James Purdey was then thirty years of age. The premises in Oxford Street and the shooting grounds in Harrow were transferred to him together with the sum of £1,600 as 'capital to enable him to carry on the said concern', on the understanding that he should repay this sum to his father in regular instalments together with any sums of money he might receive from outstanding debts owing to the business and which amounted to £9,902. It was later agreed, however, that the repayment of debts clause be waived, and James the Younger agreed to pay his father an annuity of £1,000 per annum for the remainder of his life with the added bonus that the father could take all the produce from the kitchen garden at the Harrow shooting grounds, and also cut the grass from the 6-acre plot and sell the hay. 'I am quite agreeable to execute an agreement as you propose. But as it is heavy work on my part it will be necessary for you to get some responsible Person to join you as Security for the due performance of the agreement. I will also pay half of the expense of repairing the house', wrote the father during these negotiations. His son's reply was to the effect that he could not get anyone to join him as security, but he was sure that the lawyers would make it safe enough for the father.

The arrangements were legalised in a Grant of Annuity dated 25 August 1860, made between 'James Purdey the Younger of No. 314½ Oxford Street, Gunmaker in the first party, and James Purdey the Elder, of Royal Hill Cottage, Queen's Road, Bayswater, Gentleman, of the second part'. The agreement included the assignment of the remainder of the lease of 314½ Oxford Street, which had been agreed in 1840 for a period of twenty-one years, and the remainder of the lease of the shooting grounds. In August 1860 the lease of the Oxford Street premises was renegotiated for a further period of twenty-one years as from September 1861, to expire at Michaelmas 1882. Detailed provision was made in case the son should fail

James Purdey the Younger (1828-1909) in 1858, when he took control of the firm

to stick to his side of the bargain or predecease his father, in which case the latter could take £6,000 — the value of a Government Annuity of £1,000 if purchased for the remainder of his life.

Obviously father and son also had special family reasons for this deal and the son was pushing the settlement forward as fast as possible, for the father had written him a letter on 18 July 1860 saying:

> You are quite right to wish to secure this affair as threats have been held out by some greedy and selfish part of the Family and it will be best to put it out of their power to injure you. I do not doubt your Honourable Intentions, but I could not trust some others with a quarter of the

amount. I do not like to provide a Dinner every day for anyone who chooses to come and bring who they like un-invited and hanging about all day; it is Not agreeable. I shall always be glad to see Conny [his daughter-in-law] and you as you come by, letting me know so that I can have a good dinner for you, but I don't like anyone to take the liberty to come every day un-invited.

His wife, Mary, had died in 1860, and James the Younger entered her death in the Family Bible as: 'Mary Purdey (Born July 21st 1790) died Feby 1860 at Royal Hill House, Bayswater. Mother of James Purdey. The *Kindest, dearest* friend I ever shall have; J.P. and *deeply lamented* by me. J.P.'

James the Younger had married on 19 March 1851 in St George's, Hanover Square, at the age of twenty-two, Caroline, aged nineteen, eldest daughter of George Thomas of 3 Bolton Row, 'From where *my dear wife* was married'. The couple settled in 17 Warwick Crescent. Their eldest son, and third James, was born on 11 June 1854, and christened at St Mary's Paddington. His sister Florence was born in 1855 and the second son, Athol Stuart, who was to inherit the firm and do so much for its prestige, on 27 January 1858. He was not baptised, however, until 10 March 1860 when the Reverend J. A. Foote, curate of St Mary's Paddington, was called round to Warwick Crescent to officiate at the emergency baptism of his nine months' old younger brother, Algernon George Ingle born 24 May 1859, who died immediately after the ceremony and was buried in Willesden Cemetery. Presumably both children had caught one of the then fatal childish diseases, and had to be baptised as soon as possible under the Private Baptism of Children in Houses service. It was not until ten years later, on 7 April 1870, that Athol and his younger brothers and sister — Constance Julia born 1863, Cecil Onslow born 1865, and Percy John born 1869 — were christened in public at Marylebone Church, thus fulfilling the tenets of the Church for full public baptism of infants.

After 1860 there is no further information in the letters or the ledgers as to the activities of James Purdey the founder of the firm. He spent his time either in his house — Rifle House — in Margate where his sister Martha had died in 1853, or his cottage in Bayswater appropriately named Rifle Cottage. He died in Margate on 6 November 1863 at the age of seventy-nine. 'My poor, dear Old Father' is the entry made in the Family Bible by his son; his Will, however, gives an idea of his shrewdness and his wealth.

Four trusts were set up for the benefit of his son, James, his daughters Elizabeth Field and Jane Routledge and 'My kind and faithfuly friend, Martha Hatch, Widow, and to her heirs and assignees for ever'. James was left the firm, factory and business, as well as Nos 4-5 Rifle Terrace, Queen's Road, Bayswater, a garden and brewery at Hendon; a farmhouse

Caroline Thomas, first wife of James the Younger, and mother of James III, Athol, Cecil and Percy

and lands at Lower Cumberworth, Silestone, Yorkshire; seven freehold houses on the east side of Inverness Terrace, Kensington Gardens; the stables and outhouses belonging to Royal Mill Cottage, Queen's Road, Bayswater; and two financial trusts, one for James, and the other for his children. The two daughters received the income from the trusts for themselves and their children, and Nos 1-2 and 7-8 Rifle Terrace, Bayswater, between them, including his carriages, houses, stable and horses.

Martha Hatch was also left his freehold property at Upper Clifton Place, Margate, with its stable and coach houses; Royal Hill Cottage, Bayswater; seven freehold houses on the west side of Inverness Terrace and Nos 3 and 6 Rifle Terrace, together with the lease of 8 Coburg Place, also in Bayswater. Two servants received £100 each.

GOLDEN YEARS AND NEW PREMISES

James the Younger, when he got control in 1858, had to take immediate steps to reduce the amount of debt outstanding to the business, the extent of which has already been mentioned. There was nothing of the subservient tradesman in the Purdey family; they knew only too well that unless they received the money they would go bankrupt and the craftsmen they employed would be thrown out of work. James had no intention of allowing this to happen; in fact as early as January 1856 he had signed a Letter of Licence against Thynne H. Gwynne Esq not to arrest him for his outstanding debt of £196 8s 1d until after 25 July 1857, unless default of payment be made. Poor Dr Forsyth had suffered the same difficulties, and had written to his partner, James Brougham, on 24 June 1818: 'But I hope that now when money is more plentiful, the great English Gentry who drive about in their carriages through London, Bath etc., will think it is decent, I shall not say honest, to pay tradesmen's accounts though they cannot be compelled to do so.'

Very soon, as a result of James's actions, the financial position improved. His grasp of the various stages of gunmaking was extensive, he had trained in his father's factory under his father's workmen (notably Henry Lewis, the head stocker), and he had absorbed the financial details of running the company. He was a man of tremendous energy, great financial acumen and compelling personality and charm. He was also extremely tough and overbearingly autocratic, but had the shrewdness and technical knowledge to be able to select and pay for the best craftsmen in the trade.

At that time, the Industrial Revolution was in full swing; huge fortunes were being made and those with money and country estates were vying with each other for the prestige of bigger and better bags of all types of game. As a result, the 'Golden Years' of the gun trade were beginning. The customers who ordered the guns were, however, very knowledgeable about the type of weapons and propellents they required, as can be seen from the detailed instructions given in the Order Ledgers. Competition between the various gun firms was intense and any firm who lost its reputation for reliability was very soon 'dropped' by the shooting fraternity.

The railways had made rapid travel available to great areas of England, Wales and Scotland, the new fast steamships made it possible for rich clients to visit any part of the world to shoot big game, and so orders for shotguns and rifles of infinite variety started to pour into the shops of the London gunmakers. The sport of live-pigeon shooting was growing quickly, with Monte Carlo, Madrid and Hurlingham as its centres, and the desire to excel amongst the protagonists of this sport further stimulated the demand for guns which were absolutely reliable under extreme conditions of weather and stress. Woe betide the gunmaker whose gun let down a client during the final stages of a big competition — every other competitor and spectator was watching!

New and important customers came to Oxford Street. The King of Italy bought a rifle in 1861, and in 1862 the Prince of Wales bought his first pair of Purdey double guns. The Spanish nobility, headed by the Duque de Frias who ordered in the same year, also started to come to Purdey's, and the King of Siam was to order his first gun in 1871. Large orders were also placed through State Residents for individual Indian princes during 1862 and 1863. Lady Howard de Walden, from the British Legation at Brussels, ordered a 54-gauge revolver (Adams principle) with 200 cartridges and 4lb of No 2 powder; and both Lord de Grey (later Lord Ripon) and Lord Walsingham made their first appearance in the ledgers in 1860. Walter Winans — the great American philosopher, poet, soldier, explorer and artist — bought his first rifle and double gun in 1862, and presented Purdey with a folio of his drawings of stag shooting in the Highlands, together with two signed bronzes of American cowboys sculpted by himself.

James the Younger, like his father before him, always kept his eye open for a little extra revenue and, when the Universal Private Telegraph Company — later the Postmaster of the Western District — wanted to run their cables across the roof-tops of the houses on the north side of Oxford Street, he claimed £1 rent per annum for such installation from 1862 to 1867, and thereafter £2 per annum until 1875.

He also did a lively trade in a special invention of his own known as the Vermin Asphyxiator. This machine, which was supplied with sulphur paper, smoke paper, tobacco paper and a four-foot tube for pushing down rat holes, sold for £3 11s 6d. Sir Percival Dykes, the Reverend Henry Pickering of West Chiltington, Sussex, and Lord Northwick were the first purchasers of this contraption; but after the Prince of Wales invested in one in 1873 sales increased wonderfully, and over the next few years a steady income came to the firm as a result of this idea.

Mr Pickersgill bought the first breech-loading gun with the Purdey 'slide forward action' in 1863, but subsequently trouble arose as to the effectiveness of the mechanism with foreign and English cartridges, and several guns had to be modified. In the meantime, Purdey sold his cartridge-loading

Walter Winans, American philosopher, poet, soldier, explorer and artist, who bought his first rifle and double gun in 1862

machines not only to his customers but also to the trade, Boss taking seventeen between 1862 and 1863, and Rigby three during the same period.

James also acted as an agent for shooting estates, advertising the Mar Forest on behalf of the trustees of the Earl of Fife in the *Field*, *The Times*, the *Morning Post* and *Bell's Life*, keeping a commission of £26 3s for his efforts.

As already mentioned, the live-pigeon shooting sport was booming, and on 22 December 1862 Colonel the Honourable H. Annesly ordered an extra highly finished breech-loading double gun for Frank Heathcote Esq of Southampton, with a silver pigeon fitted into the stock on both sides and engraved: 'Presented to F. Heathcote Esq., by 100 Gentlemen being the Crack Shots of England.' Extra care was taken in fitting up the presentation case. Amazingly Mr Heathcote in 1864 purchased from Purdey a patent sewing machine per request from Messrs Wheeler Wilson & Co for £15, and in 1869 his account had to be written off as a bad debt.

The majority of guns now being built were centre-fire although pin-fire and even muzzle-loaders still appear under new orders, and the first over-and-under, 16-bore rifle No 8348, sighted between 50 and 600yd, was completed for Aga Ali Shah in 1871.

The Queen ordered a 60-bore double rifle in 1867, and the Princess of Wales gave her husband a 12-bore breech-loading double gun on 9 November of that year for his twenty-sixth birthday, a gold shield being let into the stock engraved 'Albert Edward, Nov. 9th 1867'. The following year she gave him the pair to this gun for his twenty-seventh birthday, engraved in the same way, but with the date 1868, and added to this present a paid of double-barrelled pistols with silver butts engraved with a carved monogram 'A.E.' surmounted by the Prince of Wales Feathers. The guns were returned to Sandringham at the request of Mr Jackson, head keeper, on 18 February 1895, after being held in Purdey's for some years, the barrels being worn through and the guns unsafe. A brass label to this effect was attached to each gun for reasons of safety.

Purdey kept in with his most influential customers by tipping their servants in the time-honoured practice of patronage. In 1868 over £62 was paid out in this way, examples as follows:

Mr Calthorpe's Valet	£1
Earl Cowper's Valet	£1
Lord Huntingfield's Valet	£1
Mr Mildmay's Servant	£1
Mr Baring's Servant	£1
Viscount Hamilton's Valet	£1
Earl de Grey's Steward	£7
Lord Ilchester's Valet	£1
Duke of Devonshire's Valet	£1 10s

He obviously worked out the amount in proportion to each customer's social standing in the field.

But, just when everything was going so well and James was really finding his feet, family tragedy struck. On 7 January 1870 his wife Caroline suddenly died at the age of thirty-eight. 'The Dearly beloved Wife and Companion and most true and faithful Friend ever to be *Loved* and *Lamented* by me. J.P.' is the entry in the Family Bible recording this disaster, and James must have been shattered. The family had been extremely happy at 28 Devonshire Place which, by 1869, had become the new family home. Apart from this house and Rifle House, Margate, the family also enjoyed all the facilities of James's steam yacht *Lynx*, with a crew of six. He had purchased silver and china for this yacht in 1870, charging these items to the firm. Now he was left with a growing family on his hands, the youngest of whom was only a year old, and the eldest, James, fifteen.

The records show that the volume of work and orders he had to contend with was remarkably high so his time must have been fully taken up with the running of his business, his family responsibilities and also his duties to the Gunmakers' Company, of which he was a member of the Livery, and about to become an officer of the Court. Trade exhibitions were on the increase, so he also had to travel abroad on behalf of his business.

The rearing of game birds had increased the bags of pheasants and this is shown not only in the increased orders for cartridge-loading machines and the powder and shot to go into them, but also by the large numbers of filled cartridges sold to customers from 1873 onwards. The Prince of Wales ordered 6,000 cartridges for his centre-fire guns in 1873. The vogue for sawdust-filled cartridges was evident, His Royal Highness ordering 4,500 in the same year. Sawdust cartridges were made by Eley and designed to produce better patterns of shot by binding the pellets together with sand or sawdust. The results were fairly good, but occasionally the shot charge 'balled'. They were not recommended to be used in muzzle-loaders in case a small piece of hot sawdust left in the bore should ignite the powder charge during the next loading process. However, another form of cartridge was invented and was considered superior in every way. This was the wire-mesh cartridge in which the shot was enclosed and held together in a thin wire container until it was clear of the barrels. The wire then broke with the pressure to which it was subjected, and the shot expanded to its proper pattern. For long ranges a small-mesh wire cage was provided in a green cartridge, and for short ranges a wider mesh one in a red cartridge. Sir Victor Brooke of Colebrooke and Lord Malmesbury, two of the greatest sportsmen of the age, were great protagonists of this type.

However, 1870 was a watershed for the fortunes of the Purdey family, for in this year James took out the patent for his bolt and top-lever mechanism for opening and closing guns. The invention enabled the sportsman to open his gun with one easy movement of his right thumb on the top lever without changing his grip of the hand on the stock. Until now, the shooter had had to put his hand under the guard to open the gun, but now the lever was connected to the bolt by a spindle, which withdrew the bolt from the lumps in the barrels, allowing the barrels to fall down so that the spent cartridges could be extracted and new ones put in. A spring on the lever pushed the bolt back into the slots on the barrels when the gun was closed. This invention, now that the patent has run out, is the basis of the opening mechanism of all guns manufactured, but in 1870 it made Purdey's fortune and reputation certain, and revolutionised the whole concept of safety for the breech-loading gun. One of the main reasons for the invention's success was the safety factor of the grips on the barrels, and the confidence that this gave the shooting fraternity. As has already been pointed out, the more conservative shooters had doubted the efficiency of many of the methods used by individual gunmakers to ensure a tight and secure 'joint' between the barrels and the face of the action. Here, at last, was the answer.

In 1876 there was much controversy amongst the great shots of the day as to the value of various degrees of choke in guns. Lord de Grey, later Lord Ripon, used choke barrels; his great rival and namesake de Grey, later Lord Walsingham, swore that choke was unnecessary and used cylinder barrels. As a result a competition was held at the Gun Club in Notting Hill to decide between the two factions. Purdey, Lancaster, Grant and Boss guns with cylinder barrels did best. The following year, 1877, James Purdey offered a 50 guinea silver cup to be shot for at pigeon between cylinder and choke guns. On this occasion the cup was won by Mr Cholmondeley Pennell using a choked gun over all ranges from 30yd to 40yd. The controversy continues to this day.

But to return to Oxford Street. The records for 1867 give details of James's rates and taxes:

1867

Parochial Rates:	Oxford Street		2 years	£9 9s 7d
Income Tax:	"	"	2 years to 20 Sept	£2 1s 8d
House Tax:	"	"	" "	£3 2s 6d
Land Tax:	"	"	" "	£1 17s 6d
Male servants	2			£1 1s 0d
Carriages	2			£2 15s 0d
Horses	2			£1 11s 6d

These were high for the time, but fortunately for James the business entering the premises, however, was increasing out of all expectation. The Viceroy and Governor General of India, Lord Mayo, had ordered 6 rifles, 2 centre-fire guns and two pairs of pistols in 1869, and several guns were supplied to Manton's of Calcutta. New and secondhand guns continued to be sold, but a great deal of business was being done in the alteration of existing locks of guns to the rebounding principle. Previously the hammers of the locks had forced the strikers into the primers of the cartridges and held them there; this inevitably caused 'drag' when the gun was opened as the strikers tended to tear the cartridge case. The rebounding lock allowed the mainspring to push the hammer back slightly from the striker after firing, thus taking the pressure off the strikers and allowing them to withdraw from the cap of the cartridge, permitting easy opening. Lord Calthorpe had this alteration carried out to his guns in June 1872, as did Mr Hussey Vivian, the future Lord Swansea.

At the same time, breech-loading pin-fire guns continued to be converted to central fire, and a typical bill for this work, sent to both Sir Victor Brooke and Lord Kinnoull, reads as follows:

Pinholes stopped up on Barrels; Action Softened; Piece Brazed on

Detonating Nipples and Plunger Extractions; New Hammers fitted; Action and Hammers Polished; engraved, hardened and Locks converted to Re-Bounders with new Springs and Tumblers.

Nevertheless, the new gun was always the most important source of income, and 225 shotguns, 54 rifles and 2 pistols were delivered during 1873, one of the new customers being Sir Richard Wallace, founder of the Wallace Collection at Hertford House, who ordered four breech-loading guns with all extras for his illegitimate son Captain Wallace.

But of as much interest to those in the firm as this, was James Purdey's own behaviour, for in 1872 he had started smartening up his personal life. He bought an oval, covered Russian dressing case for £2 5s, and in 1873 he spent over £270 on 'doing up' his personal toilet ware; he repaired his portmanteaux, gilded a florin for £1 1s, and at the same time gilded a scarf pin. He repaired his toast-rack and bought half a dozen razors. On top of that he purchased wine coolers, gilt epergnes, a gilt rosewater dish, peppers and salts, '5 pairs of "FISH EATERS", made as new', and much else besides, as he was 'courting' and intended to marry again.

And he did marry again. On 12 August 1873, at the age of forty-five, he married Julia Haverson at St Paul's Church, Penge. She was nineteen years old. It must have been a daunting proposition for Julia to take on five children aged between nineteen and four years, but even more challenging was the prospect of having to cope with James with his charm but in other ways forbidding character and his growing business. That she was able to manage admirably is shown in the letters of real affection she wrote to her stepson Athol in later life and the devotion shown to her by the rest of the Purdey family. She was, after all, the same age as her eldest stepson, James, when she took the place of an adored mother and devoted wife.

She bore him six children, only two of whom survived after the age of eight — Sefton and Lionel Bateson. Purdey's eldest son, James, was now a young man learning the business in all its complexities and skills. He was sent to the factory to learn to build the various parts of a gun, in the tradition of his father and grandfather, and was obviously a great source of pride to the family; but he was physically weak, already beginning to show signs of the consumption which was to end his life at the age of thirty-seven. Athol, at fifteen, was still at school, but was destined, after completing his education, to join his elder brother in the firm.

In April 1874 Athol Purdey was sent to a school at 40 Rue de Lorraine in St Germain-en-Laye, run by M. la Serres. His first letter home said that he enjoyed France. Evidently the liking for his new school did not wane for in 1876 his brother James, in Paris on a business trip, was asked in a letter from his father to see how Athol was getting on and told to tell his younger brother that his father had received favourable reports on his progress. In

James the Younger in 1873, at the time of his second marriage

the same letter, dated 22 April, Jemmy, as his father called him, was told that the 'Frenchmen who want 16-bores would be persuaded to change to 12-bores if they were told that they could be made very light as we have a pair of Top Lever Bar, bottom Rib with high elevation that could come [weigh] under 6½lbs. if required'. In the letter there is also the first hint of Jemmy's health being a cause of worry. An abscess had formed on his leg which would not heal. 'Keep quiet for your leg. I hope it is no worse. Some of you write twice a day. I like it.'

Purdey was building yet another house at Margate, and Jemmy stayed with his father and stepmother at 2 Fort Crescent, Margate, while the new

Julia Haverson, second wife of James the Younger

house was being built. The rest had not helped him as an extract from a letter to Athol in France shows:

> ... as he looks thin and miserable and I am very, very anxious about him and it looks more serious than it did ... Dr. Paget said that as the change had not proved beneficial he advised cutting down to the bone to let out any matter and to take away the inflammation. He operated yesterday, Dr. Clover administering Laughing Gas, Jemmy did not know it was done — there was no dead bone, but the perioxtium [sic] was separated from the bone. They hope he will now improve.

GOLDEN YEARS AND NEW PREMISES

He urges Athol to write to his brother and

> . . . in writing do not say I am anxious but say that the operation is over and that the Doctors speak so favourably about him. I have nothing to suggest in alteration of your studies and you may learn fencing if you wish and think it serviceable.

The patient must have shown some improvement for the next letter to Athol, dated 11 May 1874, makes no reference to the operation. Purdey acknowledges two letters from his son and ticks him off about his spelling: 'There is one thing you must study very carefullyand that is your spelling, for it is very defective, and the best way is not to disguise it; ask La Serres' or Moody's advice. You likewise forget your capitals frequently.' He had bought 'a new horse (which now has a cold), is very good and matches Prince to perfection — they make a grand Phaeton pair'!

His next letter to Athol from Fort Crescent, told him that Jemmy and his father had been to Brussels on business and the wines and dinners had not been up to the standard of the Café Anglais in Paris, with the result that he had been thoroughly unwell and had pain in his kidneys and back. 'Try and have your clothes, books and furniture in order and a place for everything as you will find it a great assistance in business, my dear Boy.'

It was in this year that Purdey was elected Master of the Worshipful Company of Gunmakers for the first time. Like his father, he had been elected to the Livery, and after that to the Court in 1867, serving in that capacity until taking over the responsibilities of Renter Warden and Upper Warden. In November 1874 he travelled down to the Proof House and Court Room in the Commercial Road. 'I presided over my first Court', he wrote to Athol, ' on the 1st of the Month, Julia and Laura going with me. I got thro [sic] my duties, I think, satisfactorily'. He must have done his duties well as he was re-elected to the position of Master in 1882 and 1892, beginning the tradition of Purdey service to the Gunmakers' Company which has continued to the present day. His sons Athol and Cecil were Masters on eight occasions between them, his grandsons James and Tom, sons of Athol, on two and three occasions respectively, and Harry Lawrence, Sir Victor Seely, Major Jok Purdey, son of James Purdey IV, Ernest Lawrence, Lawrence Salter, and the author, have been Masters of the Company on nine separate occasions since 1952, making a total of twenty-five Masterships from Purdey's covering 1841 to 1983, an average of a Master every six years.

Purdey found both the finances and the Court in dreary order. Having revived the financial position he proceeded to put some life into the social side of its affairs and organised the most lavish dinners in the Court Room for the ladies of the Court Assistants. Twenty different dishes, seven different wines, musicians and singers were laid on; such entertainment being held as often as possible, sometimes seven times a year, and greatly enjoyed. His son, Athol, while unable to entertain so lavishly during his terms of office due to the financial climate at the end of World War I, also jollied up the Courts by organising a shoe of faro after the supper which was provided at the end of each meeting.

Apart from this responsibility, Purdey continued his search for perfection in his family. Poor Athol, then aged twenty-two, gets another rap over the fingers from Birmingham in 1880:

> My Dear Old Boy,
> Many thanks for your letter, still lots of faults. I have sent letter CORRECTED — you should always put capital letters at commencement of letter or sentence. You must not think me fidgetty or unkind, because you must be CURED and its best to let you know as you grow. The weather is very nasty but warmer today.

Athol must suddenly have seen red as his answer to his father has no capitals and deliberate bad spelling.

But whatever was going on in his family life, Purdey continued the expansion of his business. The foreign trade increased amazingly and between September and December 1876 the names of over seventy new customers from the continent — Belgian, French and Austrian — appear in the books. Cartridges and pigeon guns seem to be the main content of their orders, and on 8 December 1874 Mr Purdey presented a pigeon gun to the Cercle des Patineurs as a prize to be shot for in Paris. The market in the United States was beginning, and Messrs Grubb of Market Street, Philadelphia, acted as agent in that area, a discount of 20 per cent being paid to their executive, Mr Riley. A fixed scale of discounts was produced for dealers which worked on a sliding scale, as entered in the ledgers:

> Upon the first 4 Guns *12½% off* our usual price — Upon all beyond that number up to 8 inclusive 15% — Upon all beyond 8 sent me within 12 months 20% or if within 13 mos. no objection will be made to the additional month after that period and for another 12 months to start afresh at a commission of *15%* on all guns up to 8 & *20%* on all beyond 8. *These terms are strictly for payment on completion of the orders in this country.*

In 1876, the Prince of Wales was to pay an official visit to India, and suitable presents had to be found for the Indian princes. In 1875, therefore, Her Majesty's Government placed an order with Purdey's for 40-bore, 60-bore and 100-bore rifles to be presented to the Maharajah Sir Jung

Bahadoor, the Maharajah of Viziawagran, the Nawab of Iniwra and the Nizam of the Deccan. The barrels were engraved, 'H.R.H. the Prince of Wales K.G., G.C.S.I., to [name of prince]', and a gold oval in the stock was engraved with the badge of the Star of India. Mr Purdey has written alongside, 'All of these taken by H.R.H. the Prince of Wales to India as Gifts, 1875.'

In all specialised trades dealing with a small circle of customers, continuity of personal control is very important; so in 1877 James Purdey drew up a Deed of Agreement between himself and his two eldest sons, James and Athol. He intended the two boys to have a share in the business in the future, but first he wanted to familiarise his customers with the idea by 'changing the name of James Purdey in which the said Business has hitherto been carried on to "Purdey and Sons".' But it was on the distinct understanding that the change of name conferred no partnership rights on either of the boys. The business was to remain his alone, and should either son lead any customer to infer that either of them had any share in the business, or commit any act which might embarrass his father, he would have the power to remove their names without any further discussion and replace his own name as it originally stood. This document, signed by all three participants on 21 December 1877, is printed in full in Appendix 1.

The most remarkable Ledger entries at this time, apart from the continuing orders for new guns, concern those for maintenance and repairs. The Prince of Wales, Prince Duleep Singh, Walter Winans, the Dukes of Rutland, Bedford and Abercorn were consistently good customers. Charges for taking guns to and from the shooting grounds at Hurlingham predominate during the months of May, June and July, and include cab fares in bad weather, man's time for loading, cleaning the guns and supplying cartridges. Visits to Hurlingham constituted an important part of the social season, and some customers made the journey several times a week. H. Jaffrey Esq and Captain W. Duncombe were each shooting seventeen days in both June and July 1876, and Mr Aubrey Coventry attended on twenty-four days in July 1877. To save customers the trouble of taking their guns to the grounds on each occasion, Purdey let guns there for a small charge throughout the season, and detailed accounts of these transactions are entered in the ledgers. Over seventy guns thus lent in a year became quite common. The tipping of customers' valets and other servants increased substantially at this time, even more so to those considered especially deserving; for there is an entry, 'Gun to Mr Gibbs, Sir F. Johnstone's Man, a breech loading double gun by Rigby. No charge.'

The Countess of Cardigan, widow of the earl who led the Charge of the Light Brigade in 1854, ordered a pigeon gun to be delivered to her house, 115 Rue de Chaillot, Champs Elysees, Paris, during 1876 for the Comte C. de Lenderman. This remarkable woman lived on at Deene Hall, Northamptonshire until 1915, having by sheer will-power obtained her greatest wish of outliving Lord Cardigan's heir, Lord Robert Brudenell-Bruce, dying there at the age of ninety-one.

Thomas Sopwith, the engineering genius who laid out the railway companies' lines and levels, and then became the leadmining agent at Allenheads in Northumberland for Mr Wentworth Beaumont's 'richest lead mines in the world' as he termed them, bought a muzzle-loading rifle in 1876. He cannot have guessed at the surrounding moors' future reputation for grouse bags, some of the best drives of which are still called by the names used when the pack-pony teams carried the ore over the hills from Allenheads to the railhead at Allendale — Half Way, Carry Away etc. It was his grandson who won the great name for the construction of aeroplanes such as the Sopwith Pup, and racing yachts such as *Endeavour*. In the same year Lord Walsingham stipulated, most emphatically, that the trigger pulls on his guns must be exactly: Front 3½lb, Back 4¼lb. These are the standard weights used on the triggers of our guns since that time.

Captain Tryon, CB, RN, bought a new pair of centre-fire double 12-bores in 1877, but these guns had only sixteen years of use before they went down with their owner, by then Vice-Admiral Sir George Tryon, KCB, in HMS *Victoria*, after she had been rammed and sunk in the Eastern Mediterranean by her sister ship, *Camperdown*, while carrying out fleet manoeuvres on 22 June 1893. As the *Victoria* sank her beautiful wooden fittings broke loose from the decks and the men in the water, together with those in the ship's boats, were subjected to the added horror of spars and planks shooting up through the waves maiming and transfixing them as they swam or smashing through the bottom of the boats. Colonel Gascoigne ordered a double-barrelled breech loader in 1877, the right barrel of which was a 20-bore for game shot, and the left barrel a .450 rifle. The action was fitted with clip-sides to take the different pressures involved.

In 1878, the Prince of Wales started his eldest son, Prince Albert Victor, on his shooting career at the age of fourteen when he provided him with a 16-bore gun. Two years later his second son, Prince George, later King George V, started with a similar gun at the age of fifteen. Also in 1878, Queen Victoria gave a gun to the Duke of Coburg and the Prince Imperial of Austria bought a rifle. The Count de Montreuil tried to get the Purdey agency in Russia. James Purdey's note reads: 'See last paragraph but two in our letter to Lanshay as to quoting to other dealers in Moscow.' The Russian trade was booming, as was the whole European market, and a random selection from ten consecutive pages of the accounts during 1880 shows the variety of countries involved:

Bela de Gyurky from Pesth
Luc. J. Scaramanga, Taganrog; South Russia

Baron Hottinguer, Paris
The Mayor and Municipality of Dieppe (pigeon gun)
Count Eugene Kinsky, Moravia
Princess Demidoff, Florence
Baron C. Mecklenburg, Vienna
Prince Belosselske de Belozersk
Count Gyula Karolyi, Hungary
Prince Emile Furstenburg, Bohemia
Count Potocki, Poland
H.R.H. The Duke of Chartres
H.R.H. the Crown Prince of Hanover
The King of the Hellenes

The Marquess of Ripon ordered a large tin-lined case to hold the new weapons he had ordered to take to India on his appointment as Viceroy in 1880. His battery of guns included a 12-bore shotgun, a 4-bore rifle and no less than four 500-bore double Express rifles complete with ammunition. The word 'Express' had been coined by Purdey to publicise the performance of his rifles. He likened their performance to a railway train, which travelled with great velocity and a flat trajectory, and this witty description caught the popular imagination and has stuck ever since. These rifles were, of course, built centre-fire, although Lincoln Jeffries Esq took delivery of a muzzle-loading 577 rifle in 1877. However, fewer and fewer muzzle-loaders were ordered and the heavy centre-fire rifle became steadily more popular, the Duke of Portland ordering a pair of .450 Express double rifles, with Lancaster barrels and Turner locks, in 1880.

In 1878 a great exhibition was held in Paris, and Purdey decided to participate. He sent across thirty-six guns and rifles of all descriptions, and decorated his stall with the Royal Arms and the Prince of Wales feathers, two Royal Warrants and eleven photographs from the Long Room. His eldest son James was in charge, and took his own tools so that he could carry out any quick repairs to customers' guns. Out of all the guns taken only twelve were sold and these with a discount of 10 per cent; however he won an award which was proudly displayed the following year in Australia at a trade exhibition in Sydney. On that occasion a similar display was mounted and fifteen guns and rifles were sent out. The prices of the rifles were £125, and the best gun cost £66; four guns of lesser quality — C, D and E — were included. Seven guns were sold and the remainder returned to England. Purdey helped to defray costs by selling the showcase for £4. The international exhibition in Calcutta in 1883 was also something of a disaster, as only six guns out of a total of thirty-three sent to India were sold, all of these being of B and C quality; the best quality rifles and guns had to be returned to England.

James Purdey the Founder,
attributed to Sir William Beechey

James Purdey the Younger,
by Archibald Stuart-Wortley, 1891

Athol Purdey by Oswald Birley, 1927

Lord Henry Bentinck

'A Happy Family' by Sefton Purdey, 1907. Julia and her sons go for a drive

A French cartoon by R. Sala Nézière extolling the penetration qualities of Purdey cartridges. The caption was added by Tom Purdey: 'Franchement, Monsieur, qu'en pensez-vous . . . ?' 'Mes compliments, cher ami, vous avez d'excellentes cartouches, leur groupement est excellent et leur pénétration parfaite. Les Purdey, sans doute!'

The introduction of various qualities of guns had been a great mistake and recognised as such at a later date. Thinking that there was a ready market for those bearing his name, Purdey decided to build guns covering the whole price range from the best gun to the mass produced variety. He therefore imported machinery and turned out guns of A (the best) at £66, B at £55, C at £33, D at £30, and E quality at £18. Below that he made a boxlock gun of very inferior workmanship. He told his customers, and the trade at large, that all these guns were excellent value as the same finishers who completed the top-quality guns also finished the other weapons to the same standard. But this proved to be nonsense as, being made to work on lesser quality work, the standards of the finishers soon declined on the top-quality guns. Immediately the cry went up that Purdey was having his guns built in Birmingham and elsewhere, and then charging top prices on the grounds that they were hand-finished in his own factory. As a result, his customers began to think that the top-quality guns were coming down to the level of the other grades. Purdey appears to have taken some time to see the danger he was in, and the other grades of guns continued for at least another five years until only one quality, the best, was reinstated as the only Purdey-built gun.

The lease of the old Manton shop would expire in 1882 and James not only wanted a shop of his own, but had the resources to build one. His respect for the first Duke of Westminster seems to have been reciprocated, for he obtained permission to build his new shop on a site leased from the Grosvenor Estates. The Duke wished, as far as possible, to have commercial occupants for his property south of Oxford Street, and such sites were often offered to speculators. In 1879, Purdey began to buy up the subsisting short-term leases of Nos 57-60 South Audley Street with a view to rebuilding, as the occupants preferred to sell their old leases to him rather than to rebuild themselves. In the same year he submitted to the governors of the Grosvenor Estates a proposal for the reconstruction of the corner house of Mount Street and South Audley Street and the two properties to the south, having previously acquired the lease of a large draper's shop on that site. The existing buildings were old, the first ones on these sites having been leased in 1730 to the undertakers to the Grosvenor Chapel and, seemingly, most still survived. They were pulled down and the present Audley House was designed by the architect and surveyor, William Lambert, in the new Queen Anne style favoured by the Duke, and completed by the builder, B. E. Nightingale, in 1881–2.

Purdey's property speculations also included the contract for the rebuilding of 31 Green Street. He was the leaseholder of the old house on that site but, as he was required by the Estates to use the services of Sir Robert Edis as architect, he assigned his contract over to the latter. In 1884 the Duke was surprised to find Purdey preparing to complete the block up to

The architect William Lambert's drawing of Audley House, 1881. Note the warrants granted by Royal Houses, displayed on every window from the corner of Mount Street to No 60 South Audley Street

Alford Street as a speculation, but agreed he could carry out his present design better than anyone else. However, the occupying tenants of Nos 61-3 wished to carry out their own rebuilding and Lambert's designs were extended in 1889–90, the sitting tenants keeping their leases down to No 63 known as Alford House, so had James completed the contract he would have built all the houses in South Audley Street from No 57 to No 63 and the ensuing rents would have been prodigious. The year 1892 was to see Purdey building once more. He took 84 Mount Street, and once again employed Lambert to design the building of a shop, surmounted by a small flat, the building being carried out by Stanley G. Bird. This is the site of our present accessory shop.

In spite of its respectable air, South Audley Street had had its full share of disasters. In 1768 the windows of Lord Bute's house, No 75, were broken by a Wilkite mob; and in 1792, in a jolly form of tit-for-tat, Mr Wilkes had his windows broken at adjacent 35 Grosvenor Square by the Mount Street Rioters who were complaining of unemployment, their livelihood having been destroyed when their small shops and workplaces were pulled down for housing development. These windows of Mr Wilkes were very special 'parlour' windows, consisting of large panes of plate glass engraved by his daughter with eastern scenes, and recorded as being of great beauty. Thomas Goode's at 19 South Audley Street suffered when a mob of unemployed persons smashed their windows in 1886, and a letter of sympathy was sent to the owner, Mr W. G. Graelle, by Queen Victoria.

But to return to James Purdey's building project. The lease of the shop in Oxford Street was sold to the Royal Zoological Society, and he got down to his great design for Audley House. With the object of making his gun shop the most famous in the world, he spared no expense on its adornment, and the new premises created a great stir. People drove to see the new building in the course of its construction, and either eulogised over its design, or deplored its ugliness in contrast to the old Georgian houses in the vicinity. After the completion of the new building, the move took place during 1881 and an article in *Land and Water* dated 6 July 1889, gives a description of the new premises. After referring to 314½ Oxford Street as a 'dark and inconvenient place, somewhat confined in space from which fine weapons were sent out', the writer refers to the new building as a 'palace among gun manufactories, as no-one can view it without conceding that it stands without a rival in the world amongst such establishments, and not only so, but one of the handsomest buildings of any description in London'. Due to the good light and ventilation, far more of the actual gun work could be done on the premises than was done in the generality of London gun shops, and done better, since the surroundings were much more convenient and agreeable to the workmen.

The workshops themselves were in the basements and cellars. Then there were four floors above the entrance hall where the clerks, with their ledgers, sat round the walls of the front shop. The first was a flat which contained a sitting room for the Purdeys. The second and third floors were occupied by stockers, finishers and engravers, and the top floor was the cartridge-loading area. Cartridges were loaded by hand in blocks of 100, and a hand-lift was installed from the basement to take the powder and shot to the loading machines; 30lb of raw powder was licensed to be held on the premises together with 500lb of explosive made up into safety cartridges. More than 2½ million cartridges were hand-loaded each year.

The pride of Audley House was the Long Room, situated at the back. It was greatly admired and contained photographs of famous shots — Lord de Grey, Prince Duleep Singh and Mr Walter Winans — all of which are

The Long Room in 1882; the original layout after the completion of Audley House. In the foreground can be seen the stock-blanks for customers' gun stocks

still in situ, and was described as a cross between a gun room in a first-class country house and the luxurious smoking room of a London club. James Purdey had felt that this room would make an excellent picture gallery should he require extra revenue, but in the event he kept it as his personal office. Among the mounted heads and curios referred to was an American split-cane fishing rod valued at £200, the stoppers to the ferrules being made of rubies and the reel of gold. The article continued: 'The value named will not appear so "American" as the rod, and the whole affair illustrates the notion of what passes for sportsmanship with a large section of would-be American sportsmen.' A prized possession was a stuffed white pheasant which was shot and presented at Sandringham to James Purdey by the Prince of Wales in 1886. This remained in the Long Room until the 1930s.

A magnificent Royal Coat of Arms was built into the front of the building, over the main entrance, in such a fashion that it could not be removed; and to demonstrate his pride in the possession of the Royal Warrant, James surrounded the building with railings topped with Royal Crowns.

Once installed in his new and very grand shop, James started to live very well. He was concerned that his horses and carriage turn-out should be of the best and his appearance, both in London and in the shooting field, was immaculate. He lived in Devonshire Place, being driven to work every morning in his phaeton. His footmen were dressed in the smartest of liveries — red tail-coats edged with gold and with royal-blue cuffs, lace cravats, yellow waistcoats, yellow plush breeches with silk stockings and pumps. Sefton Purdey was a competent artist, but unfortunately there are no drawings of James's carriages and coachmen, although some exist of his later motor cars; but Sefton's sketches of his father's family life are charming.

Although James greatly enjoyed shooting, it was not one of his strongest skills. Shooting at Elveden in 1878 from October 15–17, as a guest of Prince Duleep Singh, the party included Lord Leicester, an excellent shot, and James the Younger, who freely admitted that while he was the best gunmaker he was also the worst shot in England; this was probably one of the reasons he was put on the flank. The entry in Lord de Grey's game book reads as follows:

Elveden October 15th–17th 1878

Partridges and Pheasants, Hares and Rabbits

Guns		15th	16th	17th
de Grey		249	119	217
Walsingham		200	108	140
Huntingfield		157	65	106
Leicester		83	85	55
Flanking	Maharajah	55	18	9
	Purdey	30	33	20

KILLED 90 PARTRIDGES IN ONE DRIVE!!! adds de Grey, underlining each word four times.

James did, however, enjoy all the good things of life and his correspondence with his customers, written in a strong hand, shows great self-confidence. Such confidence was fully justified by the perfection of manufacture of Purdey guns, but occasionally error or carelessness inevitably occurred. Earl de Grey, who had ordered five sets of three guns between 1865 and 1880, returned one set with the comment that 'surely Purdey did not expect him to pay for these'. Another customer wrote that he considered the whole firm was fraud, as in his opinion the Purdey gun was no better than anyone else's and only sold on its name — a frequent charge! An even more serious rebuke was to come from Sandringham some years later when the Prince of Wales' set of three guns was sent down as ordered, but on arrival the cases were found to be empty. The letter to Mr Purdey read as follows:

14th January, 1894

I hope these mistakes will not happen again. If it is one of your Workmen you should give him a severe talking to. Anyone ought to have been able to know that the guns were not in their cases by the weight. I hope the Prince will not hear of this mistake.

He, I am sure, would be very angry that good care was not taken. I was frightened myself that someone had taken the guns on the Road or Railway. However, I hope everything will be alright. I hope now I shall get them tomorrow morning.

Two days later Mr Purdey received acknowledgement that the guns had arrived safely.

A very straightforward attack was delivered by a Mr Frederick Beauclerk, who had tried to borrow £2 of the firm's money from Athol Purdey, and who must have received a stinging letter from the head of the firm. Mr Beauclerk's reply is as follows:

June 6th 1883. 94, Jermyn Street

Mr Purdey,
Though I should have been very glad to have had a pair of your guns as I am satisfied that they are not to be beaten in excellence of workmanship, after your letter of yesterday I decline to deal at your shop. My asking your son to lend me the paltry sum of £2 seems to have frightened you so much that you forget yourself, and insinuate that I am not a gentleman. It shall not be my fault if you do not repent your letter. I never forgive such an insult, especially from a tradesman, and I have heard lately a story of one of your guns bursting in a very dangerous manner, which, when well ventilated, as I shall take care it is, will hardly rebound to your credit or increase your custom; I wish for no impertinent rejoinder on your part; Messrs. Newburn Walters & Co. are my solicitors, if you have anything to say you can write to them.

Frederick Beauclerk.

Unfortunately for Mr Purdey, the story of the burst gun turned out to be all too true, for in January the same year, Mr Frank Shuttleworth of Hartsholme Hall, Lincoln, had written him the following letter:

1883, Jan. 4th   Hartsholme Hall, Lincoln

Sir,
I write to inform you that one of the last batch of cartridges you sent

Athol Purdey (1858-1939) as a young man

me to Old Marden (Schultze Powder) has blown my gun all to pieces. The barrels present the appearance of a pitch fork having burst asunder at the breech and divided to within 6 inches of the muzzle.

I was knocked down stunned and the gun, which I had thrown down after the explosion, was picked up in 3 pieces. I don't know how I have escaped being killed. Part of the brass end of the cartridge was taken out of a small wound in my cheek. I shall hope to call on you tomorrow about 12.30 when I shall hope to see you and give you all particulars. The thing is quite clear and that is that your cartridges are not safe.

Yours faithfully,
Frank Shuttleworth.

After what must have been a most trying interview, Mr Shuttleworth appears to have changed his mind as to the cause of the accident as the following letter explains:

> Sir,
>   I have only shown my damaged gun to my relations and one or two of my most intimate friends who are all of the opinion that the gun must be at fault, and I must confess that I quite agree with them. Under the circumstances I naturally feel no confidence in the fellow gun which was made at the same time by the same workmen whom you say you were obliged to send away for the reason you told me of, and therefore I shall never shoot with it again. Under these circumstances I think you ought to replace my guns by two new ones of rather more strength, in which case I will at once return you mine. The price you charge for guns ought to guarantee their safety, and I feel sure you will not think my suggestion at all unreasonable.

Mr Purdey, however, wrote that he did not agree with this suggestion, exciting this rejoinder from Shuttleworth:

> *Jan 17th, 1883*
>
> Sir,
>   I am sorry to say that the explanation given in your letter of 15th Jan is no more satisfactory than the long interview I had with you a day or two after the accident. I feel positive that there is no blame to be attached either to my loader or myself . . . I cannot, of course, attempt to enter into details after our long interview, but if *no* blame is to be attached to you (seeing that both the gun and the cartridges are of your make) you will not feel aggrieved if I ventilate the whole subject in the Sporting Papers in order that the matter may be thoroughly discussed. I am quite satisfied in my own mind that your gun was either defective or too weak for the cartridge you applied . . . I still think my two guns should be replaced by new ones.

Purdey's contention had been that the gun had not been properly closed when the trigger was pulled, either due to dirt in the breech, or due to the fault of the loader, but this was vigorously denied by the owner. In the end, Purdey's replaced the damaged gun only to pair with Mr Shuttleworth's surviving gun, so there must have been some truth in James's suggestion.

Besides this, suggestions and rumours that Purdey guns were built elsewhere than by Purdey's themselves were rife, though James did battle with anyone who made such accusations. In 1866 he had conducted a lively correspondence with Mr C. Reilly, Gun Manufacturer, of 502 New Oxford Street, who was alleged to have told a customer that Purdey's farmed out their guns to him to build and 'one had been proved and stamped by Purdey for which you had to pay him'. James did not accept his explanation and wrote to his informant, a Mr Bateson, saying so. He also writes:

> I have through jealousy and a bad mean spirit been sneered and slandered by two or three in my trade, but as I have not found anyone who like yourself has given me the particulars and therefore the means of convicting them of their lies, I have been obliged to treat it with the contempt it deserves. I get up only one quality of work, the best, have my *own men*, make my own guns upon *my own premises*, pay a *larger price*, and have men to *overlook* and *perfect* work in the finishing state such as are employed, I believe, nowhere else.

Several similar charges were reported to him but, being only hearsay, nothing could be done legally to stop the abuse. Times have not changed in this respect; the same charges and insinuations are made today, in almost the same way.

An even more annoying ruse was the unauthorised use of the Purdey name. James the Younger's view was that if a man was good enough to start on his own, he should do so under his own name. Unfortunately, his father had used the words 'from J. Mantons' on his original case labels, but the son had the greatest contempt for the practice. In the 1880s various craftsmen left Purdeys to set up on their own account and used the words 'from Purdey's' after their names. William Evans was one of the most famous of these and Purdey took legal advice over the matter. Counsel's opinion was that, as the words were true, nothing could be done about it. The interesting reaction was that the public and customers alike condemned this practice and letters, both signed and anonymous, arrived in the Long Room.

On 22 January 1885, T. Shilston Esq. wrote from Highweek Street in Newton Abbot, Devon:

> I have in my possession a gun: on it is engraved 'William Evans (from Purdey's)'. Would you kindly inform me what he was in your Company. I have reason to think not a practical man, but simply a shopman — if so he certainly ought not to make use of your name as it is likely to mislead the public. I should feel greatly obliged if you would give me the information.

An anonymous letter which arrived in Purdey's on 20 January 1885 expresses worry, not that Purdey's name should be hurt, but that people should be tempted to think that the other gunmaker's products were of the same quality:

Brighton
Jan 20/85

J. Purdey Esq.,
Dear Sir,
I consider it would be policy [sic] of you and a protection to your celebrated name if you were to let the public know through the 'Field' that *all* the men that leave your establishment do not leave because they are *superior workmen*, for no doubt you often discharge them because they are not. There is a man named W. Evans selling guns of very *inferior Birmingham make* and having 'from Purdey's' engraved on them and scores of gentlemen are led to believe by him that he was one of your leading workmen and by his advertisements etc., they think he turns out the same gun as you do at half or one third the price, you ought to send someone there to hear the yarn he spins about your guns and his. The cause of my writing to you, two young friends were led away by his advertisement and bought guns.

A letter to the press on this subject by a correspondent signing himself West-end Gunmaker was answered in a very indignant letter to the *Field* the following week from Frederick Beesley, inventor of the Purdey Action and now in business on his own account. On 24 January he wrote:

Sir,
Your correspondent 'West-end Gunmaker' has done good service by his remarks on the sale of cheap guns with assumed names; but his remarks on makers stating they are 'from' some well-known firm are at once indecent and uncalled for.

Apart from the fact that persons such as he describes would not be likely to acquire sufficient means to start in business for themselves, buyers can always ascertain what has been the connection with the 'well-known firms'.

In many instances, as in my own, documentary evidence can be produced showing that the services rendered to those firms have been retained and appreciated during some years; and it must be admitted that it is by the technical knowledge and skill of those who work for them that the 'well-known firms' sustain their well-deserved reputations.

Your correspondent is not justified in making such aspersion under a nom de plume; and I trust to your well-known fairness and sense of justice to insert this reply of mine, as I claim to be fully justified in the form of my advertisements in your paper and otherwise.
Fredck. Beesley
(Late on the staff of James Purdey and Sons)

(We have received letters from other makers to the same effect. — Ed.)

Although nothing legally could be done, the matter certainly caused Purdey a great deal of annoyance and worry, and considerable expense from obtaining legal advice as to the possibilities for protecting his name.

But such embarrassing and annoying happenings apart, business during the 1880s prospered. Queen Isabella II of Spain had a gun sent to 18 Avenue Kleber in Paris in 1881 and the King of Spain ordered a 12-bore gun in the same year. With the ever-increasing popularity of pigeon-shooting competitions, James gave a 12-bore pigeon gun as a prize to the Antwerp Pigeon Club in 1881 as an extra form of publicity on the continent. The Duke of Croy gave Archduke Frederick of Austria a gun in 1882. Mr P. Wigram, on the other hand, would be obliged by Messrs Purdey & Sons not sending him any more circulars 'as at present each one cost him a penny postage as he is no longer at Icklingham'. James continued to lend rifles and guns for the shooting season, the two sons of the Prince of Wales — Prince Edward and Prince George — borrowing deer rifles for three months during 1881–2. The Prince of Wales himself bought a new set of three hammerless breech-loading guns in 1885 (Nos 12087, 8, 9), and the Sultan of Turkey ordered a 12-bore gun in 1886.

Although the hammerless gun became more popular, several prominent sportsmen such as the Duke of York, later King George V, Lord Ripon and Lord Walsingham continued to use hammer guns for reasons of safety, claiming that it was possible to see at once if the hammers had been drawn back to the cocked position, and also that the hammers gave a better sighting plane down the rib, so preventing your cheek coming too far over the stock.

In 1885 James again gave a Purdey gun as a prize to the Cercle des Patineurs of Paris (see overleaf).

So popular had game shooting become that special ledger pages for orders for cartridges were introduced from 1885 onwards. Until then, orders for the comparatively limited numbers of cartridges were included in the ordinary customer entries, but when individuals began to order 20,000 to 30,000 each year, the ledgers had to be reorganised. Lord Sefton ordered a steady 16,000 cartridges annually; and Lord Aveland, at Normanton Park, ordered 24,000 in 1885 and 29,000 in 1886. It was also in 1886, in September, that the first game licence was bought by Purdey's for a customer, the Honorable W. Warren Vernon.

As almost every sportsman had his own particular ideas on cartridge loading, details were kept showing individual shot charge and powder requirements, and James would discuss these with each customer before the latter's cartridges were loaded for the coming season. The powders used were black powder, EC and Schultze. In 1889, 460,000 cartridges were loaded with black powder, the number falling to 34,000 in 1898. During the same period cartridges loaded with Schultze increased from 1.4 million

Avis.

Cercle des Patineurs. Tir aux Pigeons.

SECRÉTARIAT — PARIS

1885.

Le Mardi 16 et Mercredi 17 Juin, à 3 heures précises,

The Great Champion Prize.

Un fusil offert par Mrs James Purdey et Sons, 57 et 58, South Audley Street, Londres.

30 Pigeons, 28 Mètres.

La poule sera tirée en deux jours

   15 pigeons le Mardi 16 Juin.

   15 pigeons le Mercredi 17 Juin.

Ont seuls le droit d'y prendre part, les Membres du Cercle des Patineurs.

Ont gagné ce prix les années précédentes.

- En 1872, le Comte O. de Montesquiou
- 1873, le Comte du Lau d'Allemans.
- 1874, le Vicomte de Martel de Janville.
- 1875, le Cap^e Lane.
- 1876, le Comte du Lau d'Allemans.
- 1877, Arundell Yeo.
- 1878, le Vicomte de Quelen.
- 1879, le Duc de Riansares.
- 1880, Vansittart.
- 1881, Lafond Joseph.
- 1882, le Vte de Quelen.
- 1883, de St Clair.
- 1884, Lafond Joseph.

Prix de Consolation. Samedi 20 Juin.

per annum to nearly 2 million while EC powder, after becoming popular for some years, fell off nearly to its 1889 level (in thousands):

Year	Black Powder	Schultze	EC	Total including Various
1889	460	1,396	284	2,237
1890	363	1,544	430	2,423
1891	327	1,462	522	2,414
1892	248	1,687	427	2,415
1893	213	1,885	460	2,635
1894	163	1,912	480	2,632
1895	107	2,087	412	2,715
1896	82	2,174	312	2,759
1897	54	1,950	227	2,526
1898	34	1,813	329	2,415

Elaborate safety precautions had to be taken in all such establishments under the Explosives Act of 1875, but in spite of this a private action was brought against a Mr Cadwell, a firework manufacturer, in 1889, debarring him from keeping explosives on his premises as the danger threatened the plaintiff's property. Cadwell lost his case and this judgement by Mr Justice Kekewich threatened the cartridge-making activities of all gunmakers. Fortunately the judgement was reversed on appeal in 1892. Mr Cadwell's expenses were paid by subscription from various firms interested in holding powder on their premises.

Sir Ralph Payne-Gallwey, one of the great amateur experts on shooting, whose books were widely read, bought a game diary in 1883 for Thirkleby Park at Thirsk, and a pair of hammerless guns with Third grip and Whitworth steel barrels in 1887. The following year he tried to interest Purdey in buying his patent game scorer, writing from 40 Cadogan Place on 3 June:

Dear Mr Purdey,

It would give me *much* pleasure if you could come and lunch with me here at a ¼ to two on Wednesday, and we could have a cigar and a talk afterwards. If you are, as I fear, too busy to do so, I will call at about 12 o'clock.

It is more in Belgium, America and Germany I imagine a *good* marker would take than in England. The foreign shooter is very fond of a *Dodge* in the shooting line.

I also fancy set in the Butt of a billiard cue it would take; — possibly also in a Cricket Bat. This counter is so *neat* it might be applied in many ways. If taken up it should be shown in the Paris X.

Yours truly, R. P. Gallwey.

Sir Ralph Payne-Gallwey, one of the leading gun experts of the late nineteenth century

James did not take up this kind offer as the following letter from Sir Ralph shows:

Thirkleby Park, Thirsk
October 28th, '88

Dear Sir,
    You will know by now that Messrs. Brazier and I are jointly bringing out the game scorer — *Entre Nous* — put my name on the ones you use (alone) if you so wish it: Thus: Sir R. Payne-Gallwey's Patent Scorer:

but generally the scorer will be known as the 'Gallwey Brazier'. This is only fair to Messrs. Brazier as they have improved my original Pattern.

The supply of dogs to his customers continued. In 1881 shooting dogs were sent to Spain through the offices of a Mr Briddon. And, using Mr Briddon as before, dogs were sent out to Abraham Pasha in Constantinople during 1883, complete with biscuits and straw for the journey.

The work that was now coming in at an even faster rate, was mainly due to the invention and application of the new Purdey Action on the self-opening principle which had been invented by Frederick Beesley, and offered to James Purdey in the year 1879. Beesley, writing from 22 Queen Street, Edgware, on 18 December 1879, says:

Sir,
  Having invented a Hammerless gun which I believe to be equal to, if not superior to, anything of its kind yet produced; I am desirous of meeting with a purchaser of the right to the same. It is on a principle *entirely different* to any other in the market, and also possesses a peculiar advantage, as any old gun may be converted to a hammerless one at moderate expense. I offer it to your notice first in the trade, and should esteem the favour of a personal interview, if worth your attention, when I can submit a working conversion.
                    I beg to remain,
                        Your obedient Servant
                            Fredck. Beesley.

This invention was based on the outward look of the old Purdey Action, but incorporated the means of compressing the mainsprings when the gun was closed, the old action having been based on the Anson Deely principle of compressing the mainsprings when the gun was opened. It was achieved by fitting cams into the slots at the knuckle end of the flats of the action, using the crosspin for their axis. Two $1/8$in diameter holes were drilled through the body of the action to incorporate rods which were forced back by the cams when the gun was closed, thus compressing the mainsprings. When the gun was opened after firing, the mainsprings pushed on the rods and cams which forced open the barrels, thus making the gun a true 'self-opening' gun.

Purdey immediately realised the value of this invention, one of the extra benefits being that the mainsprings were always 'at rest' when the barrels were dismantled from the gun and the gun was in its case. He therefore saw Beesley, and on 29 July 1880 drew up an agreement with him for the use of this action. This agreement was signed by both parties at 66 Chancery Lane on 5 August 1880 and can be summarised as follows.

GOLDEN YEARS AND NEW PREMISES

On 3 January 1880, Beesley had taken out a patent No 31 granting him, his heirs and successors etc, the sale and exclusive licence to make and mend in the United Kingdom of Great Britain, Ireland, the Channel Islands and the Isle of Man an 'Invention in the construction of Break-Down Guns' for the term of fourteen years provided he filed in the Great Seal Patent Office, within six calendar months, a proper specification describing the invention. This Beesley did and, in accordance with a previous agreement, he then signed with Purdey the letters patent, subject to the payment of £20 — which was paid on the signing of the documents and receipted by Beesley — and to the payment of royalties. The royalties were that Purdey was to pay Beesley 'Five shillings for every gun made by the said James Purdey according to the said Invention and Letters Patent until the number of Guns made by James Purdey, or his Agent or Agents, shall have amounted to Two hundred, whereupon such Royalty shall cease'. It was further agreed that Purdey should have the right within four months from the execution of this assignment to commute the 5s per gun royalty 'for the payment of a sum of money amounting to Thirty five pounds', to be paid to Beesley within the said four months; and if such payment were made, James Purdey would be discharged from all claims Beesley might have on him. On 16 November 1880, Beesley signed a receipt for a cheque of 'Thirty Five pounds, being the last payment mentioned in the within Assignment, and in communication for all Royalties payable thereunder'. Purdey had complete control of the new action, and protection of manufacture for fourteen years.

By 1889, therefore, gunmaking was going from strength to strength. In February 1889 James had so many orders on hand that he simply closed the firm's order books until the middle of September, although naturally he made a double rifle ordered by the Prince of Wales for Prince Eddy. The profits for the three years 1887–9 averaged £9,000. All this money went to James himself and, taking into consideration that the maximum tax that he paid on these sums was $1\frac{1}{2}$ per cent, an approximate net equivalent in modern money terms would be £150,000 per annum. One can see that he was justifiably proud of his position therefore, when in 1890 he wrote to his son James then abroad for his health: 'Lord Fife told Athol I was the most popular man he knew, everyone likes me (Gammon), and when he called on me he wanted to know what I did to make the Prince shoot so well as everyone spoke of it.'

'The Grand Old Man' by Sefton Purdey, 1907. James the Younger on his 79th birthday

THE GRAND OLD MAN

Audley House, painted in 1974 by Eileen Smith

LATE VICTORIAN AND EDWARDIAN SPLENDOUR

The years 1890–1910, when rich new customers visited the premises in South Audley Street, started with personal tragedy for the Purdey family. James Purdey III, the eldest son, had now become very ill with consumption. He had married in 1884 and a daughter, Dorothy Irene had been born in 1886, but so serious was his condition that his father came to the conclusion that the only hope was a sea voyage and a long stay in a dry warm climate. It was decided that Bloemfontein in the Orange Free State was the ideal place, and so James and his wife, leaving their daughter behind, had been sent off to South Africa in the ss *Mexican* from Southampton in July 1888.

The last letter from James to his brother Athol before leaving England, was written from Plymouth thanking him for a farewell telegram and saying that the trip down the Solent had done him so much good that he felt better already and had eaten two good meals, which was a rarity for him. They were sailing on to Lisbon. He enclosed a private code so that he could write secretly to his brother, which seems odd for a married man then aged thirty-four. The journey went off safely and they arrived in Cape Town, where they settled in an hotel before completing the move to Bloemfontein in October with the assistance of £70 telegraphed to him by his father. He wrote to his brother Athol on 26 October 1888 from the Free State Hotel, in Bloemfontein, announcing their safe arrival and news of their travels. He hopes that Athol and his father had not had too much trouble with the new Ejectors, and not too many disagreeable mishaps with the guns. His father had written to him there on 18 October, saying 'I hope soon to hear you have arrived' and immediately protesting at his asking for £70 'and you must remember that in future telegrams for money will not be attended to, and you must arrange accordingly, as I must not be put to any unnecessary expense'. The son must find employment for, 'If you improve as I believe and hope, you will have to stay out some time to make a perfect cure [and] you must not expect to live the life of a Gentleman without doing anything, or you will find it will not be good for you in any sense.' He then says that he intends bringing his third son, Cecil, into the business in order to supervise the shooting of the rifles and guns: 'The shooting etc., will want re-arranging as it has been done very slovenly.' Cecil had never intended to go into the family business. He had been placed as an apprentice to Penn's the shipbuilders, who advised Purdey on matters concerning his yacht *Lynx*. Cecil loved this work, but on his return to the family firm put all his skills and energies into the boring of barrels and rifle-making with significant effect. After telling James this news his father goes on to say, 'I need hardly say this arrangement will not injure you in the end in any way.' The letter continues by saying that it has been a bad partridge year, but a very good grouse one in which Lord Walsingham had killed 1,075 to his own gun in one day. Lord Fife had sent him a haunch of venison, the Duke of Richmond a haunch and a salmon, Mrs Jameson was delighted with her new guns and Lord Windsor had ordered a new set of three. 'Give my best love to Nell, tell her I will look after her Bairn and she must look after mine and not let him get into mischief. Your Affectionate Father, James Purdey.'

He was always writing to his son with advice and admonitions: 'You have good views and ideas, and if you had only used your capabilities, stuck to it and had energy without getting among the sharp-know-everything division, who are no good to anyone, you would have done well and not be as you are.' His son must find the cleverest men round him and listen to their views: 'how about Australia, Tasmania or New Zealand for the future, are the gold mines worth thinking about as a career?' The main thing was to get his health back and then: 'Keep away until you are well; have you thought of Reporting?' Again the warning of frivolity: 'Keep clear of the "Amusing Young Fraternity", and the *Old Wags* and sort of Companions you had.' A letter of 16 November 1888 says that they will send out the child only when the parents are finally settled as it would be very expensive and 'perhaps the Nurse would get married and leave you — and then where would you be'. And in the same letter, an astonishing remark: 'I think you have powers adapted for better things than gunmaking if you will only cultivate your talents.' This to the eldest son, trained to the business, confirmed as his heir, and virtually dying, must have been very confusing.

However, all was not just lecturing. He was contemplating buying a new shooting ground at Elstree of fifty-two acres, (four acres for building), which would only cost £4,000 and he could build a cottage on it. And 'when I am dead and all the others tired, it would be valuable for building ground'. Purdeys had beaten Cogswells very considerably at cartridge sales and the talk was that Cogswells had gone off a bit. Athol, Cecil, Percy and Carl [his brother-in-law, Carl Svenson] had dined with him at the Gunmakers' Hall, and 'I sang them many songs'. This must have been an entertaining evening as Mr Purdey was extremely musical and sang well. The food was also of high quality and he loved good wine.

James Purdey III (1854-90), the eldest son of James the Younger

The son's letters home describe the various English people he met and interesting happenings, particularly the official and unofficial goings-on at the inauguration of Judge Reitz as the new president of the Orange Free State. Mounted burghers rode in from all over the state, bringing with them their wives, children and servants in wagons. The burghers were all armed, many with muzzle-loaders, and on arrival in Bloemfontein were issued with 1lb of powder and a 2s ticket for grub. After that they loaded their muskets and fired off salutes as ordered by their commandants from 4.15am until 6am. Hundreds kept arriving and firing volleys, and the noise was fearful. A huge procession formed up at 9.30am consisting of the Volksraad (Parliament), judges, clergy, government officials, army, police, mounted burghers still firing, and all the 'rag, tag and bob tail of Bloemfontein, with the new President wearing a low felt hat and three other pals in an old 'C' spring open carriage', and proceeded to the Dutch Church for the official swearing in. After this was over the procession re-formed and marched to the Presidency where all the guests were supposed to shake hands with their host. 'However, the Boers and their Fraus made straight for the liquor and grub. The men collared the champagne and the women lifted up the fronts of their dresses, filled them with food, after which they went into any room in the house and had a picnic.'

This letter caused great amusement in the Purdey household, but also concern for James's health as he admits that it has taken him several hours to write because of his having to stop and rest many times. It ends with love to everyone, thanks for looking after the baby at Christmas and also 'Love to the Governor [his father] — Military Governor I think he ought to be called now'. The Governor had written several times urging him to get well and almost ordering him to find work in the Transvaal, which he hears is the coming place for money making. In November 1888 he mentions he had altered the Long Room by having the staircase shortened and moved to where the fishing rods were stacked for sale, and had lost the chance of continuing the building of all the houses from No 57 to No 63 South Audley Street. He then apologises for always giving James advice but he misses him so much, especially as he considers James has so many more advantages than his brothers in every way and if only he could get well there should be a great future for him. Yet when his son asks for extra money he is told to be more careful as his father is also paying for James's baby, nanny, the servants and the upkeep of the house in Putney. At the same time he blames himself for his son's faults which are forgiven and forgotten for ever more from this time on and will not be again referred to but adds, 'you know a good bit of your life has been wasted as you had excellent abilities.' He was always in his family's thoughts and minds, they all missed him terribly, and the Duke of Alba had asked how he was improving.

By early 1889 it looks as if, behind the scenes, Nell was writing a much more realistic type of letter home without James's knowledge, as his father's letters become more gentle and conciliatory, even telling him to do exactly what the doctor orders and to do 'nothing in the shape of work or to tire yourself'. But James was dying, and his family must have been told of the position either by Nell or the local doctor.

A shoot at Eridge Castle, Sussex, on 19 November 1889. From left to right, standing: General Bateson; Mr Akers-Douglas MP; HRH The Duke of Cambridge; Lord Randolph Churchill MP; Lt-Col A. C. Fitzgeorge; Sir George Wombwell Bart. Sitting: HSH Prince Victor of Hohenlohe; Maj-Gen Sir George Harman

However, dying or not, his father could still get back to old form very quickly where expense was concerned. A letter written at that time pointed out that his father's expenses in keeping his son and his household amounted to £686 19s 8d for six and a half months; and it was far too much. He had further told the chief manager of the Bloemfontein bank to restrict his monthly payments to £43, and he and Nell must keep within that limit. Just to cheer up poor James after this bombshell he adds that Julia not only has gallstones, but also has developed a kidney complaint, osteo-arthritis and a weak heart; her new doctor, Garrard, is a silly old woman, and she is very queer and very weakly. The adored baby girl [Dorothy Irene] has turned out to be jealous with 'a pugilistic turn', and has developed bad habits from being in the company of servants all the time. She should not be kept from her mother for another six months.

Nell must have been appalled to read such a rigmarole and this is probably the reason that the couple, despite all their fears for James's health, decided to return to England as soon as possible.

James and Nell, therefore, packed up all their belongings, paid their bills, closed the account at the bank and set off for Cape Town. Sailing in the *Mexican,* the ship in which they had originally travelled out, they arrived home on 4 June to the delight of their whole family. But James never recovered strength, and on 9 September, 1890 he died and was buried in Paddington Cemetery. Although anticipated, it was a very great blow. James had been trained as the next 'boss'; Athol now had to step into that very responsible position.

But, this tragedy apart, the year 1890 was a very successful one for the firm. Although the 1883 Calcutta exhibition had been a failure, individual orders from the East continued to increase. In 1890 Amin us Sultan, Grand Vizier to the Shah of Persia, ordered .577 rifles for his own use and also for the Shah; Captain Douglas Haig wrote from the 7th Hussar Camp in the Deccan, ordering various accessories for his guns — the first entry we have for the future field marshal — and in November the same year the 'Ladies of Ireland' contributed enough to buy a new pair of 12-bore guns (Nos 13632/3) and present them to the 6th Marquess of Londonderry at the end of his term of office as Lord Lieutenant of Ireland.

Sir Edward Grey, later the Liberal Foreign Secretary and Viscount Grey of Fallodon, became a customer in 1890, as did Prince Henry of Battenberg and R. H. R. Remington-Wilson Esq of Broomhead Hall, Sheffield, who was given a new set of three 12-bore guns (Nos 14580-2) in return for his older set of 13000 numbers. This straight swop was made by Purdey in consideration of Mr Remington-Wilson's well-known influence in shooting circles.

The Reverend Nasson Clark of Yendon Vicarage near Leeds, must have fallen well short of the standards expected from wearers of the cloth when he bought a gun for £71 8s. 'BAD' writes Purdey, 'Would not Trust. CAUTION.' Apparently care had also to be exercised when dealing with customers who were wards in Chancery. A. F. Basset was one in 1890–1 when he ordered 16-bore cartridges, and the address of a solicitor in Cheapside with the note 'A Ward in Chancery. Use caution.' is inscribed beside his entry.

Special hammer guns were built for the King of Italy in 1890, and further new customers were the Crown Prince of Portugal, the Infante Don Antonio of Orleans; Mr R. Middleton, stud groom to the Duke of Alba at El Campo in Spain; and the Duke of Teck of White Lodge, Richmond, father of the future Queen Mary. Prince Victor Duleep Singh ordered a set of three 12-bores in 1892 and a further set of four 12-bores in 1896. The Duke of Buccleuch ordered two pairs of 12-bores for his sons Lord Henry and Lord George Montagu Douglas Scott, but Lady Randolph Churchill who had ordered cartridges in 1891 has the warning after her name, 'CARE.LONG. VERY'. The Marquis de Tanlay ordered a 7in badger trap, and books on shooting and big-game hunting were sold from the accessory side of the business.

LATE VICTORIAN AND EDWARDIAN SPLENDOUR

Athol and Cecil were now taking the brunt of all the management, with the 'Old Man' very much in overall charge. In 1891, Archibald Stewart-Wortley, a great sportsman and friend of the Purdeys, painted the famous portrait 'James Purdey the Younger' which was shown at the Royal Academy that year and which hangs in the Long Room. It shows him standing four-square in a brown tweed suit holding one of his guns and staring at the world through his monocle. Stewart-Wortley's other great picture of Dr W. G. Grace hangs in the museum at Lord's cricket ground. James had great affection and respect for Grace and actually gave him a 12-bore gun (No 14731) in August 1896. Grace used it until his death in 1915,

Dr W. G. Grace shooting in 1896 with the gun given to him by James the Younger

when the firm bought it back. And whenever Grace wanted to shoot with two guns, Purdey lent him another. He did this in 1909, 1913 and 1914. On 7 August 1895, Surrey County Cricket Club bought a pair of 12-bore best ejector guns for £182 12s as a retirement present for the secretary, John Shuter.

Cricket Purdey admired, but he did not, however, care for racing. Fred Archer, the famous jockey, had received a 12-bore gun (No 12063) in 1885 but not as a gift from James; it was bought for him by Captain Machell, the trainer. The gun, which is still in use today, was used by Archer until his death in 1886. After that it came back to the firm and was given to Sir Gordon Richards by Tom Purdey in 1933, the year in which Sir Gordon made racing history by riding 259 winners thus beating the seasonal record of 246 set up by Fred Archer in 1885. The gun has been used every season since 1964 by Major David Swannell, Secretary to the National Horseracing Museum and Jockey Club handicapper. A framed photo-copy of the account is displayed in the museum on York racecourse.

Several women customers patronised the shop in the 1890s, such as Violet, Lady Beaumont, of Carlton Towers; the Duchess of Abercorn, who bought a .577 rifle; Lady Rothschild; Mrs Campbell of Pembridge Square; Mrs Burns of Copt Hall in Essex who bought two .300 bore rifles and a 12-bore shotgun; and HRH the Comtesse de Paris, whose account was sent to her at Stowe in Buckinghamshire in 1898. Over 7,000 customers provided James with an excellent income, among them the new diamond kings from South Africa, Julius Wernher, Abe Bailey and Alfred Beit; the American millionaires, H. P. Whitney of J. S. Morgan and W. W. Astor were only two of the new American customers from almost every state in the union. In almost every case the Americans wanted inlays of gold, heavy chasing and engraving, in fact all the adornments which Purdey felt spoilt the look of his guns and gave an impression of vulgarity. However, if that was what they wanted, he provided it for them; English and continental customers however continued to order good plain or lightly chased guns. Brookes Brothers of 1 Whittington Avenue, EC were agents for the US, and it is interesting to see that the prices charged for the cases of the guns was nearly half that of the guns themselves — namely £89 for the gun, £49 for the case and fittings. Other customers came from Canada and, from Australia, a steady flow of orders with one constant customer, Sir Rupert Clarke of Melbourne whose family still have close links with the firm, leading the way.

As the 1890s progressed, the list of continental customers was a succession of illustrious names: the Duke of Braganza; Count Herbert Bismarck; the Archduke Franz Ferdinand of Austria; the Prince de Wagram, descendant of one of Napoleon's great marshals; the Duc de Gramont; the Counts Kinsky and Metternick; HRH the Prince de Ligne; King Milan of

A shoot at The Grange in Hampshire, c1892. From left to right: Lord de Ramsey; General Bateson; Lord Newport; Lord Walsingham; Lord Ashburton; A. Stuart Wortley (portrait painter of James the Younger)

Serbia; and General A. de Lignières, Commandant of the 6th Division of Cavalry at Lyon, who ordered eleven guns and rifles between 1891 and 1894.

In England the Duke of York, the future King George V, was presented with a pair of 12-bore hammer guns (Nos 14770-1) in 1893 by the borough of King's Lynn in Norfolk, and His Royal Highness made up a set of three by ordering one more gun on his own account. The guns bumped his fingers so he took delivery of three dozen finger guards, which were obviously necessary as he shot off 13,000 cartridges in that year. Rifle covers were also supplied in the Balmoral tartan. Mr L. D'Oyley Carte of 4 Adelphi Terrace became a customer, as did J. Bruce Ismay of 10 Water Street, Liverpool, who ordered a pair of guns in 1894. Mr Ismay was to become managing director and president of the White Star line, owners of the *Titanic*, built by the firm of Harland & Wolff. He travelled in the liner on her maiden voyage from Southampton to New York in 1912, and was one of the few survivors.

At the same time as Mr Ismay was ordering his guns, Hugh Gore Esq returned his repair bill for work carried out on his gun No 13227 with the following comment: 'Sirs, I had to send my gun to another gunmaker as I told you there had been nothing done by you to make it work perfectly so consider you have no claim on me.' The letter-heading of the bill shows that Purdey at that time built guns for the following royal personages: the Queen, the Prince of Wales, the Duke of York, the Duke of Cambridge, the Emperor of Russia, the Emperor of Germany, the King of Italy, the King of Spain, the King of Portugal and Archduke Franz Ferdinand of Austria.

The Honourable Harry Stonor, later Sir Harry and equerry to Queen Mary, one of the brilliant and most elegant shots of all time, became a customer in 1895. Such a reputation for style did he enjoy that men and women would stand respectfully at a distance, in all weathers, simply to see him shoot. The same applied to Lord de Grey when he shot at his home, Studley Royal, near Ripon. There, the last pheasant stand before lunch was alongside the great east front of Fountains Abbey. All the morning drives would have been leading up to this one point, and the pheasants, beautifully driven and controlled, were eventually concentrated at the 'flushing point' beside the great tower. Lord de Grey, with his loaders and using three guns, would take his place on the grass, with his back to the water-gardens, the spectators behind him. The head keeper would then go forward alone, very slowly, so that the birds would rise in small groups and

The Marquis of Ripon's (Lord de Grey) game scores, 1867-1923

Game killed by the 2nd Marquis of Ripon from 1867 to 1923

Date	Grouse	Partridges	Pheasants	Woodcock	Snipe	Wild Duck	Black Game	Capercaillie	Hares	Rabbits	Various	Total
1914	2,385	1,628	4,434	6	7	42	–	–	178	709	78	9,467
1915	3,078	2,576	2,598	17	6	5	–	–	341	594	96	9,311
1916	3,435	613	875	8	3	–	–	–	116	474	105	5,629
1917	2,087	1,159	1,990	15	4	9	1	–	168	584	36	6,053
1918	1,445	878	1,279	10	3	1	–	–	128	564	184	4,492
1919	1,097	1,151	1,185	9	–	13	–	–	156	619	262	4,492
1920	765	685	1,527	16	6	19	–	–	144	899	111	4,172
1921	1,984	1,242	2,081	16	7	18	–	–	190	793	82	6,413
1922	992	1,307	2,289	7	4	10	–	–	182	438	134	5,363
1923*	915	356	–	2	4	–	–	–	51	346	200	1,874
1867/1913	18,183 / 79,320	11,595 / 112,598	18,258 / 222,976	106 / 2,454	44 / 2,882	117 / 3,452	1 / 94	– / 45	1,654 / 30,280	6,020 / 34,118	1,288 / 11,328	57,266 / 499,547
Total	97,503	124,193	241,234	2,560	2,926	3,569	95	45	31,934	40,138	12,616	556,813

* This is only up to 22nd September 1923. On that date Lord Ripon was shooting on Dallowgill Moor near Ripon and killed 165 Grouse and one Snipe - At 3.15 p.m. after a drive in which he had killed 51 Grouse - Lord Ripon dropped dead in the heather. He was born 29th January 1852 and was therefore in his 72nd year.

not be panicked into taking off in one great flush. Lord de Grey, using all three guns, would put on a display that frequently drew subdued applause from his audience, such as when he had six birds dead in the air at the same moment, one pheasant just shot, four falling and one just about to hit the ground. If there was time, he would acknowledge the applause with a grave bow in the direction of the crowd. Should he miss two birds in succession, perhaps due to a change of wind which made the birds take a different line of flight, he never worried. Handing his gun to a loader, he would calmly sit down on his shooting stick, watch the birds' flight with great concentration and, when he had worked out the new angle, would then start shooting once more. His phenomenal speed and accuracy were partly due to his cartridges. He had been accustomed to using black powder but, during 1894, had written to Purdey telling him that he was delighted with the Schultze powder cartridges he had sent him to try at Studley Royal:

I find I shoot at least 30 per cent quicker with them than with Black (powder). I am not quite so accurate *as yet* but I make up and more by the number of shots got off. My bags of grouse on August 12 — 650, August 14 — 500 [to his own gun]. Please send me 2000 more.

Mr Oswald Birley, the famous portrait painter, became a customer in 1897 with a small order for finger guards. In 1927 he was to paint the portrait of Athol which now hangs in the Long Room. Athol wrote to thank him and received a charming letter in reply:

Dear Mr Purdey,
    Very many thanks for your writing me such a charming letter about your portrait. I appreciate it very much and can only say what a pleasure it is to me to feel that you are all so satisfied with my effort.
    The P.M.'s [Stanley Baldwin's] portrait has gone well.
                    Yours sincerely,
                    Oswald Birley

Also at that time Jack Seely, the first Lord Mottistone, bought his first gun. He later won the DSO in the South African War while serving with the Imperial Yeomanry, and my great uncle Jack then enjoyed, and I mean the word in its true sense, a distinguished political career culminating in the responsibilities of Secretary of State for War in 1912.

He resigned that post in 1914 after a conference held on 20 March in Dublin by the Commander-in-Chief Sir Arthur Paget, where it was revealed that the situation in Ulster was serious. The Liberal Government's Irish Home Rule Bill had aroused great fears and it was possible that the army might be needed. All officers were to be given the choice of taking part in such operations or not; but if they refused they would be dismissed from the army. Later that day the commander and 60 out of 72 officers of the 3rd Cavalry Brigade at the Curragh stated that they would prefer dismissal. The government were bitterly criticised for giving the impression that they intended to use the army to suppress political dissent. In the upshot there were no resignations in the cavalry division, no dismissals and no operations against Ulster. Jack Seely did resign, however, but he rejoined the army in August 1914, becoming a brigadier in command of the Canadian Cavalry Division and served with great gallantry at the second battle of Ypres. Completely fearless himself, he wrote a book entitled *Fear and be Slain*, and is credited with having put forward his groom/chauffeur for the Victoria Cross 'as he has accompanied me everywhere'. Throughout the war he rode a charger called Warrior, and so fond was he of this horse that he wrote a book about its life called *My Horse Warrior*. This was a great success when it was published, and later the horse was put out to grass in the Isle of Wight. As the years went by Warrior got older and older and more and more decrepit. After an excellent lunch one day, Uncle Jack, in splendid form, took Hugh Seely, his nephew and eventual owner of Purdey's down to the paddock to show him the horse. A little man in a bowler hat was looking over the gate. Uncle Jack, in full flood, gave the world at large a short history of the wonders of the horse, its courage at carrying him through the battles of the war, and then turning to the stranger said, 'Are you interested in horses Sir?' 'I am from the RSPCA', said the little man in the bowler hat. 'Oh!' said Uncle Jack, very taken aback. 'Oh indeed! Well, good day to you. Come Hugh, we must be getting on.' And whisked Hugh Seely away to show him the kitchen garden.

In 1895 there was another good flush of clergymen — the Reverend V. R. Carter of Alwen Rectory, Bradfield; Canon J. M. Eley of Newstead, Manchester; the Reverend G. W. Corbet of Sandown Castle, Shrewsbury; and the Reverend W. G. Weddall of Linby Rectory, Nottinghamshire. Francis Pym Esq, of The Hazells, Sandy, Bedfordshire, became a customer, as did Winston Churchill who wanted his rifle regulated and the great Cecil Rhodes, who bought shotgun cartridges in 1900.

In 1897, the Tzar of Russia, Nicholas II, ordered a pair of 12-bore guns, Nos 16006-7, through Prince D. B. Galitzine at the Gatchina Palace in St Petersburg; and in October 1900 he bought a pair of 16-bore guns, Nos 16934-5. The brutal murder of this charming man and his family at Ekaterinburg in 1918 shocked the civilised world. His guns however are better cared for, as Harry Lawrence was to discover on his visit to Moscow as a member of an Advisory Committee on Proof during 1975. At the conference, he was told by the Russian Minister of the Interior that the whereabouts of every Purdey gun was known — in Moscow, Leningrad and

James the Younger in 1895, aged 67

elsewhere. The new rulers of Russia in fact had, by then, become customers, Nikita Khrushchev having ordered four guns at different times in the 1960s for duck shooting and game, and Mr Kosygin a single gun during his term of office. Tzar Nicholas's uncle, the Grand Duke Nicholas Michailovitch, had Purdey guns sent out to him at Tigris in the Caucasus in 1897; and King Alexander of Serbia ordered 6,000 rounds of .300 rifle ammunition to be sent to the Palais Royal in Belgrade.

Debts and outstanding accounts continued to embarrass the firm but in nothing like the way they had done in the past. In fact some parents actually helped the tradesman against the behaviour of their sons, as can be seen

> 5, Bank Buildings, E.C.
> 8th November 1894
>
> Dear Sirs
>
> We beg to send to you a Cheque for £7.6.8. in settlement of your account against the Earl of Burford, up to the end of September 1893. His Grace the Duke of St Albans refuses to pay the Earl of Burford's accounts after that date and you will, if you please, understand that you must look to the Earl of Burford, & to him alone, for payment for any goods supplied or to be supplied by you to his Lordship as from that date.
>
> We are Dear Sirs
> Yours faithfully
> Flashfield & Williams
>
> Messrs James Purdey & Sons.

from the communication issued by the Duke of St Albans.

A new solicitor was helping Mr Purdey by 1897 — W. H. Herbert, Esq, of Cork Street. Plenty of commissions came his way, one of which was to help trace such customers who did not pay their bills. The Mutual Communication Society for the Protection of Trade had been established in 1803, and its secretary, W. P. Nobbs, wrote on 24 May 1904 about the whereabouts of A. Senior Smith, who owed Purdey's £100, and who was last heard of in Cape Town:

> We have seventeen hundred Smiths on our books, who are engaged in one form of business or another, but we can find no such person as A. Senior Smith. Questions have been put to Smiths whose first name commenced with an A, in the hope that their middle name might be Senior. However, this plan has failed, as well as a thorough search which was made through the principal Directories of our American Cities.

In 1897 Queen Victoria celebrated her Diamond Jubilee. The whole of London was decorated for the occasion and James, whose loyalty to the Royal Family knew no bounds, was determined to put on as good a display as anyone. He therefore commissioned from Messrs Freeman & Co of Great Chapel Street a magnificent coloured crystal illumination which was erected over the Coat of Arms surmounting the main door of the shop. Topped by a crystal crown 4ft in height, a huge glass centre-piece 5ft across by 7ft high bore the insignia 'V R', and two 6ft glass wings, shaped as roses and thistles, supported the whole display. The whole was held in position by chains and wires attached to brackets let into the brickwork and illuminated by gas jets which, when lit, caused the whole to shimmer and flicker. Flags and red, white and blue ribbons, together with placards bearing loyal greetings, were draped and fixed to all the balconies of the building, and such was its success that this display has been carefully kept ever since, being erected over the front face of Purdey's on every Coronation or Jubilee since that date, the insignia of the monarch being changed by moving the glass initials on the centrepiece. At the Coronation of Her Majesty Queen Elizabeth II in 1953, however, Tom Purdey dispensed with the wings and simply hung the crown and centrepiece on a stout cable from under the second-floor balcony, having done away with the gas jets and substituted eighty electric bulbs in their place. In this precarious position the full weight was taken by two brackets, and this wonderful decoration dangled for three whole weeks, keeping Tom awake in his flat on the first floor by banging backwards and forwards when the wind blew down Mount Street, and constituting a potential death threat to any customer entering the front shop on a windy day.

But while all this success was displayed on the business side, Purdey domestic affairs were not quite so happy. In December 1890, three months after the death of James, Athol, at the age of thirty-two, had married Mabel Field, daughter of Allan Field of 4 Upper Avenue Road, NW, and the sixth-generation great grand-daughter of Oliver Cromwell. On their engagement in the July of that year, Athol had signed a letter stating that in the case of his dying before his marriage, he left all he possessed or was interested in, with the exception of his tradesmen's debts and family jewellery, to Mabel as a token of his love and esteem. He requested his father 'my kindest friend', to see that these wishes were carried out.

Before this Athol had enjoyed life as a bachelor to the full. His visits to Paris and his description of a stag party in Brighton reveal the ebullience of his nature, and charming mementoes of his earlier romances remain. Jessie, surname unknown, of Upper Phillimore Gardens, Kensington, sent him several letters and flower-printed valentines. One letter, signed with initials only, encloses a lace-rimmed card showing blue and white violets and the message 'Hommage d'une Amitié Sincère', from F.J.V.J. 'To you there are many like me, but, to me, none like you. That is what I feel.' Whether Athol felt like this is doubtful, because a new world opens up in Dublin from 1882 to 1885, with a New Year greetings card arriving on each 31 December. There is also a card showing a butterfly about to fly into a spider's web with the warning 'Take care' written beneath, and a very saucy postcard of a fox with a guitar kneeling before an owl which is peering out of a hole in an oak tree, entitled 'Romeo and Juliet'. The fox has a dead goose hanging out of his coat-tail pocket. Mabel was not the kind to put up with this sort of thing, and soon left her mark on both Athol and the Purdey family.

Athol Purdey was very popular with the men who worked for him. He did not die until 1939, and thus there are still many connected with the firm who remember him with great affection and respect, for they know that he was a great gunmaker and also understood the business side of the trade better than anyone. Not that he did not have some of his father's temper; his rages were famous. If a man in the factory made a mistake and 'spoiled' the action or stock, or whatever it was he was making, he was sent across to the front shop with the offending piece to show it to the 'Governor', who would then decide whether the craftsman should pay for it himself — which could mean losing up to three weeks' wages — or whether the firm would bear the loss. In the case of frequent mistakes the man would be sacked. These visits were dreaded by the men as Athol would work himself up into a terrible rage and start throwing things about in the Long Room, where these awesome interviews took place.

Mabel, on the other hand, always gave the impression that she considered trade much beneath her dignity, summarily dismissing the employees of the firm from her presence with a wave of her hand, and being

Mabel Field at the time of her marriage to Athol in 1890

so unbearable to her personal servants that she dismissed maids at the rate of four a week until the employment agency, run by Mrs Hunt, told Athol that his 'Good Lady had outrun the Constable', and she would not send round anyone else for interviews.

These insights into her behaviour are disturbing enough, but since her wish that her sons be fully accepted into a very rich society had to be based on the trading capability of a gun shop, either the shop had to produce excellent and increasing profits or else the forward funds necessary to enable it to continue its activities would have to be milked. But for the present, old Purdey kept a tight control over what went on in the firm and no daughter-in-law dared to cross him.

James Purdey the Younger celebrated his seventieth birthday in 1898 and, as a commemoration, every member of the staff, factory and shooting grounds was presented with a silver match-case engraved with the Governor's initials and the date. Presumably dinners and receptions were held, as his son Athol was to hold a great dinner when he reached the same age; but there is no record of any. However, we do know that he had lost no vigour at that age, and in fact had become even more dynamic and difficult to be with.

So, unfortunately, had Mabel. After the birth of James (the fourth) in 1891, Mabel gave birth to Thomas Donald Stewart in 1897. The birth was so difficult that she nearly died, and for the next two years spent most of her time on a water-bed or in a wheelchair as her spine had been damaged. She 'took against' her husband as a result, and Tom became the apple of her eye. From that moment on, poor Athol's life became extremely difficult, for on one side he had Mabel, and on the other his stepmother Julia who, partly because of ill-health, and partly due to the feverish energy of her husband, found the kind and courteous Athol a prop on which to lean. Mabel, in her illness after the traumatic experience of Tom's birth, was ordered by her doctors to try sea air for her recovery. She therefore moved her household to Folkestone, taking with her the two boys — Jim and Tom, nanny and an under nurse, a cook, housemaid, a trained nurse for herself and a parrot called Polly. They settled in 27 Cheriton Gardens. Athol stayed in London and came down as often as possible.

No sooner had she settled in, on 9 August 1898, than the complaints to Athol started to arrive: 'I have never seen anything so awful as the people are down here . . . swarms of second rate Jews.' She goes on to a long detailed description of the local 'quality', then starts on her own household. The cook was a bad cook (she gave notice the following week), the housemaid was useless, and the trained nurse not only knew nothing about bed-baths, but had very bad breath. Rows with landladies over dogs, birds etc. were frequent, so in the space of four years in spite of her almost complete immobility, the entire party settled for various lengths of time in 27 Cheriton Gardens, 68 Bouverie Road West, 7 Trinity Crescent, Burlington House Hotel, 6 Canfield Gardens and 2 Mill Field, all in Folkestone. And in all the complaining letters she wrote, in only one is there a note of something apart from her own troubles, when she is genuinely concerned about Sefton Purdey, Julia's eldest son, serving in South Africa.

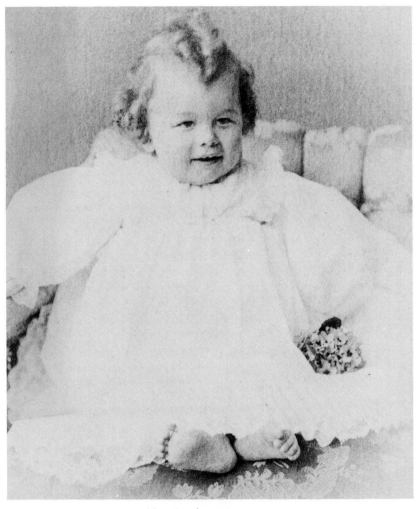

Tom Purdey (1897-1957)

Sefton, having joined the 18th Hussars, had become Galloper to General Brocklehurst in the cavalry camp at Ladysmith, after the latter had taken over command of the Cavalry Brigade from General French. He wrote to Athol from Ladysmith on 17 November 1899:

> We have had several small fights and one or two big ones. In my first engagement I was hit by splinters of a shell and a bullet grazed my left knee tearing my breeches, but I soon got all right again. They are plucky devils these Boers and seem to mean fighting. They have got us in on all sides but we can hold out all right for a long time yet.

Mabel with her eldest son, Jim, on the seafront at Folkestone in 1899

... and with her husband Athol on the same occasion

He continues that the relief column is not expected for several days; in fact it did not arrive until 28 February the following year. He wrote again on 18 March 1900 from a convent where he was convalescing after an attack of enteric fever, which had laid him low for fifty-six days, during forty-nine of which he had not been able to touch a scrap of food. He was being sent home, much against his will as he wanted to remain with his regiment. He then commented on the soft-nose Mauser bullet which flattens out like a mushroom and stops a man at once, whereas the hard-nose bullet can go clean through a man, and you could hit him several times before he dropped.

On 1 March 1900, Mabel writes to Athol by nickname from 7 Trinity Crescent, Folkstone:

Dearest Toby, [Dr] Lewis flew in this morning to say Ladysmith is relieved and being a few minutes early I hadn't had your wire. But it came whilst he was here and so I could tell him who had relieved it! Great and Good news isn't it! And all Folkestone has gone quite cracked over it. The concert is in full swing upstairs: *Comb* obligato by Jim and dances by Tommy. I am looking forward to your Birthday.

However, letters of this nature are very rare, the vast majority containing endless dreary complaints about her illnesses, her endless squabbles with the servants and her alleged lack of money. Athol must have been driven demented by the querulous, snobbish waterfall of mail which, day after day, reached him from Folkestone between August 1898 and December 1901. Even on a trip to Monte Carlo with her mother, and Constance and Carl Svedberg, Athol's sister and brother-in-law, her attacks are maintained. Athol thinks of buying a house in Hampstead and sends her the details. 'What on earth do you want to send me a budget of rot like yesterday', she replies. 'Do you think for a moment I want to wade through 14 pages of utter twaddle. I never was more angry. Please do not do it again. I have always loathed Hampstead and shall continue to do so and I am perfectly certain it makes me rheumatic, and voilà tout.' It must have been a happy family affair for she then rounds on her sister-in-law: 'If ever anyone will end by going out of her mind it is Connie. She really alarms me she is so mad and she is so awful to Carl it makes one feel quite uncomfortable. She *never* leaves him alone — nagging and quarrelling, and Carl so good to her.' Connie writes a charming letter to her brother saying how much she is enjoying herself, but pleased that she and Carl are soon going on to Rome.

That Mabel was very ill is confirmed by a letter from 25 Weymouth Street, Portland Place, signed by a specialist, A. H. Tubby and dated 1898. The treatment was that she should remain recumbent the whole day with the exception of half an hour in the morning and the same in the afternoon, and when it was hot in London she should move to cooler areas. The 'Special Apparatus' should be worn at all times. This treatment would make most women irritable, but one cannot help feel that Mabel laid it on a little too thickly. The crowning humiliation for Athol must have been the letter he received from Dr Greenwood of 19 St John's Wood Park, dated 29 July 1900.

> Dear Purdey, I am afraid you have had a good deal of trouble this winter. I heard on my return or soon after that doctors, nurses etc, etc were in attendance at your house so I did not call as I supposed that Mrs Purdey had decided to change her medical attendant. Whilst I would always rather patients changed immediately they lost confidence it would be nicer and more congenial to one's feelings if the decision came direct from the patient or those connected with them than from outsiders, especially as in your case where friendly relationships had existed for so many years. However, one must realise that trouble often clouds one's finer feelings. I sincerely hope that Mrs Purdey is now progressing favourably. With kind regards, etc.

Besides letters from Mabel, Julia wrote to Athol constantly telling him all the news about his father and, during the trying time of the Boer War, confiding her fears about her son, Sefton. Julia was also an invalid, so poor Athol had to put up with her infirmities as well as those of his own wife. Possibly hearing about Mabel's constant complaints about maids, she writes to Athol in October 1898 from the family house at 28 Devonshire Place:

> I hope, Dear, that you will find Mabel still improving in her walking powers next Saturday. You will I know be glad to hear I am the possessor of an excellent maid if she will only continue as she has started. I only wish Mabel could have found one like her — so strong and capable and such a kind old Thing — and she will massage my legs: I am indeed most fortunate.

Shortly afterwards she moved down to the Clifton Hotel at Margate with James the Younger with her. He was now seventy years old; she was forty-four. She was in a bed chair; he was in a bath-chair. In order to get some rest she recruited Flo, her sister-in-law Florence Green, to take 'Jim' out in his chair down to the jetty and then give him lunch.

> I thought it was very kind of her, but she looked tired out — he is so very restless and won't stop anywhere for five minutes . . . I keep giving Jim

ten shillings of small silver as I find when his purse is empty. Then he fidgets for Tillson [the Purdey secretary]. He will buy 6d cigars and I was awfully amused at Flo this morning. He would go into the centre of the Jetty for refreshments. She marched in with him and Jim asked for a soda and whiskey and Flo said a large soda and told the 'Gov' it was too strong. Consequently poor old Flo got more than she bargained for, I told her that he was used to his whiskey in the morning.

The 'Old Man' wrote to Athol from both Margate and Folkestone. At Margate he had seen Charles Eley, his cartridge-making friend, looking very well and much younger than himself although four years older. The jetty had been washed away during a storm, and his rheumatism was awful. From Folkestone the news was that the old harbour was being filled in and a new one was being dug out.

In 1900 Julia writes that her husband continues to give cause for anxiety:

I asked Dr. Nichol to test his Urine for me as he kept Trotting off to his Bedroom every half hour; but after minute examination Nichol said there was nothing wrong with his Bladder, in fact for his age he was one of the strongest men they ever met, and one of those men who will live for another twenty years, and if he would only be more moderate in his whiskey his memory would improve.

He was very jolly at Christmas, wearing a paper crown all evening, nevertheless his mind was wandering and when Julia was back in Margate he stayed in London. His servants kept her informed of his goings on and she writes: 'He goes to bed sometimes at 7.30–8 and is up and downstairs at 6am to South Audley Street at 8.30 and says the men are lazy devils not to be there sooner.' This letter followed one telling Athol that the Governor had been down to see her but was so restless and tiresome that she had to send him home and it was much better that he was back in London. Her kindness and tact come out in a letter to Athol when Mabel must have been very trying and Athol was thinking of moving into Gloucester Terrace. She had persuaded the 'Gov' that it was a good idea, and if Athol felt he could not meet the expense 'I am sure, Dear, the Gov. will help as it will add to your comfort in more ways than one, and Jim will do anything to make you more comfortable in your home life — and Mabel can get out in the Park quite easily.' Nevertheless, the 'Gov' was bored, and a week at the Oatlands Park Hotel, where they drove for two hours amongst the pine trees and then sat under the trees until dinner, ended in disaster as 'he got so restless he does not know what to do with himself, as he doesn't read or take any interest in anything', and she had great trouble in keeping him there.

James the Younger in the drawing-room of his house in Devonshire Place c1900. From left to right: Percy; Mrs Blandford (Julia's sister); James the Younger; Julia Purdey

James's behaviour in the shop was even more eccentric. He used to pop out into the front shop and look round. 'Clark', he would say to the head shopman, 'You are fired!' and pop back into the Long Room. Half an hour later he would go out again. 'Clark! Why haven't you gone? I fired you!' 'I am just packing up sir', would be Charlie Clark's reply. Ten minutes later the Governor would come out again: 'Very well! Don't you ever do it again!' Clark of course had never done anything at all. The staff never knew where they stood except in one respect, namely never to talk to the 'boss' after lunch as he was often so tight that you were likely to be sacked on the spot.

But in spite of all his domestic difficulties, Athol ran the business well. In August 1899 the Duchess of York, the future Queen Mary, gave her husband a double-barrelled .303 rifle engraved with his name; and the Princess of Wales gave a pair of second-hand shotguns to her brother, Prince Charles of Denmark. Lord Fincastle of the 16th Lancers, eldest son of the Earl of Dunmore, who had won the Victoria Cross with the Malakand Field Force in 1897 and was to be awarded the DSO in World War I, appears in the books. But Captain J. Edwards Heathcote, from Newcastle-under-Lyme, complained that his barrels were rusted between the tubes due to bad welding when he last sent in his guns for repair, and did not see why he should settle his account. This complaint was sweetened by a charmingly polite letter (shown overleaf) written to Athol from the 5½ year-old Prince Edward of York, later King Edward VIII, thanking him for making him his first gun, a .32 garden gun.

> 24th December, 1899
> York Cottage,
> Sandringham.
>
> Dear Mr Purdey,
> Thank you very much for the nice gun you sent me. Papa is keeping it for me. I wish you a very Happy New Year.
> from
> Edward.

He had addressed the envelope himself to 'Athol Purdey Esq', which caused Mabel to write to Athol from Folkestone on 28 December:

> Dear Old Thing,
> Thank you for your letter and enclosure. It was very nice of the little Prince to write and it was most unusual that they addressed you by your Christian Name. How did they know it I wonder.

The King of Portugal ordered yet another pair of guns in 1900, but these were to have no engraving whatever, and the actions were to be blued instead of hardened. In August the same year, Athol patented his own invention of a special shooting glove. The Office for Patents accepted the design and added: 'There will be no objection to your sending a pair of gloves to the Duke of York and it is not necessary to stamp on the articles the word *Patent*.'

In 1900 the lease of Avery Row ran out, so the whole factory moved to new leasehold premises at 2 Irongate Wharf, off Praed Street, in 1901. This building which was rented and converted to factory requirements, had been the warehouse and dwelling of a chinaware merchant.

On 22 January 1901, Queen Victoria died at Osborne House on the Isle of Wight. She was eighty-two and had reigned for sixty-three years and seven months, during which time the country had enjoyed a period of

York Cottage.
Sandringham.
Norfolk.

26th December

Dear Mr Purdey

Thank you very much for the nice gun you sent me, Papa is keeping it for me

I wish you a very Happy New Year from

Edward

peace and prosperity so great that inflation hardly existed. The price of Purdey's guns when she came to the throne in 1837, was £55 each. When she died the price of a gun with equivalent workmanship was £84. It had not doubled in sixty-three years.

The entries for the Prince of Wales in the books suddenly cease in mid January 1901, and the heading 'H.M. The King' is substituted. From the point of view of the business these years were to be some of the busiest and most profitable the firm has ever known. Shooting parties while Edward was Prince of Wales were one thing, now that he was King they became splendid in their luxury and the number of birds reared and shot. The great landowners and shooting rich vied with each other to invite the King to shoots where birds were in greater and greater abundance. The King, of course, was very circumspect as to whom he shot with, but that in no way diminished the enthusiasm of the rest.

Another great change was taking place — motor power was taking over from horse power, and the new King was a keen motorist. Many makes of car came on the scene, and in 1904 the Honourable Charles Rolls with his partner, Mr Royce, produced their first motor car to the highest standards of mechanical design and passenger comfort. Three generations of Purdeys had been doing just that in the gun trade for over ninety years, and so we have always felt that the maxim 'Purdeys are the Rolls-Royces of the gun trade' is the wrong way round. In our view, Rolls-Royces are the Purdeys of the new car trade.

While the new reign opened up even brighter prospects for the Purdey family, it unfortunately began on a sour note for the firm when, at a private dinner party, allegations were made that nearly led to a breach between Audley House and the well known and highly respected firm of Birmingham gunmakers, William Powell. The letters speak for themselves, and the sensitivity felt by the Purdeys at any slur on their name or product is obvious.

Audley House
57/58 South Audley St.
W.1.
Nov. 12, 1901

Dear Mr Powell,
    A matter has been brought to our notice by an old customer, which requires clearing up.
    He tells us that at dinner the other evening where there was more than one of our customers present, he heard it stated by a guest that he had been informed at Powell's that they were very busy and could not undertake anything at present, as they were busy on Purdey's work, having had an order for two dozen guns from Purdey's, and they must get them out of hand.

Now, as not only is there absolutely no truth in such a statement and, as far as we know, not the slightest ground for the fabrication of such a story, we are going to write to our customer, telling him he was quite right in contradicting it, as he did.

We can but feel convinced that nothing said at your establishment could by any twisting have been converted into such a statement, and I feel sure you will be pleased to have the opportunity of denying it, and your denial we shall ask our customer to bring to the notice of the gentleman who made the statement, and of those to whom he made it.

I was not aware you manufactured guns except, like ourselves, to sell retail, and though we know your guns, and fully appreciate their merits, we have never had the slightest business dealings.

I have a pleasant recollection of meeting you last June, when I was down with the other members of the London Proof Committee, to confer with the Committee of the Guardians, and as soon as I heard the story, I said I should write to you personally.

In the position we hold in the trade we hear all sorts of extraordinary statements from the public, but when such a direct and untrue statement is made, the matter cannot be passed over.

Yours faithfully,
(Signed) Athol Purdey.

The 'Great Four' c1900. Standing: Prince Duleep Singh; from left to right, sitting: Lord Huntingfield; Lord de Grey; Lord Walsingham

(Rec'd. Nov. 14 1901)

35, Carr Lane,
Birmingham.
13 November 1901

Dear Mr Purdey,

I am glad you have written to me on the subject mentioned in your letter, and afforded me the opportunity of contradicting the absurd statement made in presence of your customer on the occasion you refer to, there is no word of truth in it in any shape or form.

I cannot conceive how any man could have invented such a cock and bull story, unless it was done by way of a joke — such a fabrication seems to me damaging to the reputation of my firm as well as yours, evidently we are too busy to accept orders, which we should be glad of, owing to the large trade we are doing with guns.

I have been asked before now, 'Who makes Purdeys guns?' 'Where are they made?' 'Do you make them?' And have always given the same answer. 'Purdey makes his own and is a bona fide gunmaker in the strict sense of the words.'

I could not say more or less, and have never given expression to any word here or elsewhere that could convey a contrary meaning.

We do not make guns for the trade and have no desire or intention of so doing, and it is a retail business like yours and has been so for 3 generations. If you can find out the name and address of the gentleman who made the extraordinary statement, I will write and tell him we have done no business with you, and ask him for his authority.

I was pleased to make your acquaintance on the occasion of your visit to the Proof House, and am only sorry that anyone should have attempted to disturb the good feeling which exists between us, and will I hope.

Yours very truly
Wm. L. Powell.

Athol Purdey Esq.

Little Acton,
Wrexham.

Dear Sir,

I showed the enclosed letters on Thursday night, and as usual in these cases, the men who had said it in the first instance had not a word to say, and were completely shut up. Anyway, it has convinced all those that were present that there was not a word of truth in the statement and they said that if they ever heard it said again they would know how to contradict it. I am very glad Powell put it so strongly. My brother is coming up shortly with a gentleman to order 3 new guns from you.

Yours very truly,
Sydney M. Crosfield.

35, Carrs Lane,
Birmingham,
29 Nov. 1901

Dear Mr Purdey,

I am obliged by your letter and glad to know that your customer who heard the absurd statement did not believe in the truth of it. I should have been surprised if they had come to any other conclusion. I should very much like to know the name of the gentleman who made the statement, and if he happens to be one of our customers to ask him his authority or grounds for making it, if an opportunity presents itself of your obtaining this and will let me know, I shall be glad — I do not consider we have got to the bottom of it yet, and to my thinking it wants clearing up.

Yours very truly,
Wm. L. Powell

Athol Purdey, Esq.

Added to this, Percy Purdey was in financial trouble. He had started a business on the Stock Exchange — Purdey and Wilson, 10 Throgmorton Avenue, with the extraordinary telegraphic address of 'Heartiness'. In February 1900, he had written to Athol asking for a loan of £10 to pay the rates as a previous loan had to be spent on gas and water rates. Then, in 1902, he and his wife had to leave their London house and go into furnished rooms at Hastings as his firm had gone bankrupt. After talks with his partner, he decided to call a meeting of his creditors, and was in such a terrible state of nerves over it all that he felt he should stay by the seaside as, he said, he would do more harm than good if he was in the office. The Governor was most reluctant to help him out, but Julia, behind the scenes and with the help of Athol, persuaded the 'Old Man' to come to his son's assistance. On 23 May Julia wrote to Athol from Devonshire Place:

Dear Old Athol,

Many thanks for your wire — I waited in for it as I was a bit nervous the Gov. would not come up to scratch — as he seemed so bitter — however I *am glad,* as the Poor Boy's mind (according to his letter to me this morning) must be greatly relieved — and really Athol I think that after his past experience, when he once can get a start — he will succeed. I was obliged to give up Percy's letter to the Gov, which he took over to the business and has not brought back, so if you see it about would you mind destroying it as I don't want it read by others. With kindest Love. Aff'n yours Julia.

A further letter written at 10 o'clock at night continues:

My dear Athol,

After a long argument I think and hope the Gov. realises that it is wiser to help Percy now, than let him fail and tonight he has left me saying 'Well, I shall see what Athol thinks of it in the morning', and I told him we had already talked the matter over — however tomorrow I shall again broach the subject and tell him it must be decided one way or the other tomorrow and you will see the Solicitor so that the money really goes for what it is intended. His Conditions — and also if his furniture can be made over to the Governor — to have it so no one else can take it — wire, Dear, to me if he consents. Just say 'Allright'. I shall understand if he won't — say 'No', as I promised to let the Boy know and besides, I feel somehow this ought to be done to give him a last chance — so if the Gov. won't I will — I have more jewels than I ever wear and can spare a bit to give him 700 — so be sure to wire as I am most anxious he should have *this* chance. Yours Affly. Julia.

A charming and most understanding stepmother, and a model to many others in that difficult position! To the 'Gov' public failure was an anathema and it took Julia to put the family in the right perspective. Percy eventually abandoned a financial career and turned his talents to the stage, becoming theatre manager of The Grand, Wolverhampton.

After sorting out his younger brother's affairs, Athol set about the task of educating his two sons. The elder, James, or Jim as he was known, was now nine years old and a suitable preparatory school had to be selected. Mabel had put her own views on this problem in a letter to Athol in 1899. She had been recommended Praetoria House School at West Folkestone, but had reservations about this establishment; she preferred either Brown's at Eastbourne, or Hawtrey's on the Isle of Thanet. For her sons she wanted, she wrote: 'Plenty of sport and good style. I like a boy to be smart'. However, she must have changed her mind with regard to Praetoria House, for Jim was sent there for the summer term of 1901. Mr Roderick was the headmaster and board and tuition amounted to £28 per term.

Jim had, from his earliest days, shown promise at drawing. As early as March 1899, Athol had sent a few of his sketches to Ronald Grey at the Cheyne Studios in Cheyne Row for his opinion. Mr Grey thought that Jim had talent and advised Athol to let his son continue in his own way and warned him not to let his school make him draw 'pretty' pictures. As a result, drawing and painting materials show largely on the 'extras' account, along with blazers, straw hats, cricket bats, song books, galoshes, a compass and a pocket comb. The food cannot have been of the best quality as the matron bought cascara for the poor little boy every term along with quinine and toothpowder. However frequent entertainments were laid on, such as a lecture on the Chinese Empire at a cost of 2s per head; and the

Jim Purdey (1891-1963) in his Praetoria House School uniform, 1901

whole school was taken on an outing to view the fleet at Dover in the summer term of 1903. The school had a private boat for the boys' enjoyment and riding lessons were provided.

Jim did well, and as a result of extra lessons in the classics during his last year, passed his examination into Winchester College, arriving there in September 1905. His tutor, J. A. Fort, wrote to Athol in October 1905, sending his first report: 'He seems a capital boy with plenty of energy and a nice open manner. I have been much pleased with what I have seen of him.' The headmaster adds: 'Did excellent work in history in my exam and is as intelligent and sensible as anyone in the Div. He ought to have a very good chance of "Books".' At Winchester 'Raising Books' is to come top of one's form, or division and so the headmaster, at that time the Reverend H. M. Burge, DD, must have had high hopes for Jim's future school career.

LATE VICTORIAN AND EDWARDIAN SPLENDOUR

In 1909 Jim left Winchester and on 21 January 1910 he joined the Royal Military College at Sandhurst. Mabel was asked to send £75 as a contribution for the half-yearly residence, and £35 was placed to his account in the contingency fund for uniform, books, boots etc. The commandant considered that £6 for the first month of each term and £3 for each subsequent month was quite sufficient pocket money, and would enable the cadet's expenditure to be controlled.

Tom, Jim's younger brother, went to West Down preparatory school. His Diary covering this period says he was sorry to leave but that he never got on with his headmaster who never liked him. He entered Winchester in 1910 but, to his relief, was transferred to Eton in 1912. He, too, was to go to Sandhurst and gain his commission very early in the war and before he was eighteen.

But this is to anticipate events and returning to the business side of the firm itself, in 1901 agencies had been set up in Paris where Monsieur A. Guimond took an average of four guns per year and the Purdey gun, offered as a prize to the Cercle des Patineurs in 1885, kept Purdey's name in front of the pigeon-shooting fraternity. In May 1903, King Alphonso XIII of Spain, having ordered another pair of 12-bores, and a pair of 16-bore hammerless guns, approached Athol Purdey for help in founding a pigeon shooting club in Spain. Athol got in touch with the London Sporting Park and they gave advice to the King on the practicality and technique of such a

King Alphonso XIII of Spain, shooting with three Purdey guns in 1907

project. In the following year, His Majesty bought the appliances for the school for £62 15s and included an order for yet another pair of 12-bores. The King, who was a marvellous shot, used hammerless guns for game shooting but hammer guns for pigeon competitions. Athol put up a hammer gun as a prize in 1922 for the Tir au Pigeon in San Sebastian. This gun was built exactly to the King's measurements, the barrels of the gun were of the tightest choke possible and the gun number was 22200. The King entered for the competition and won hands down. Having done so, he told Athol that he had been on the point of ordering a new pair of guns but now only wanted a single gun to make up the pair! However, he did order a 12-bore gun for his son, the Prince of the Astorias.

In England live pigeon competitions continued, Athol's supplier of the live birds being E. & H. Roberts, Purveyors of Pigeons for Shooting, 167 & 50 Great Dover Street, London SE. The Duke of Alba ordered live pigeons at £3 4s for 4 dozen, and the shooting grounds also bought 6 dozen starlings for £1 4s and 16 dozen pigeons for £16 8s. Today we would think this sport heartless in the extreme, as did Charles Dickens in 1865 when he described the feelings of Poll Sweedlepipe, the barber in *Martin Chuzzlewit* who also supplied birds for this purpose:

> He [Sweedlepipe] had a tender heart, too, for, when he had a good commission to provide three or four score sparrows for a shooting match, he would observe in a compassionate tone how singular it was that sparrows should have been made expressly for such purposes. The question whether men were made to shoot them never entered into Poll's philosophy.

Such shooting matches continued at Hurlingham, Ranelagh and Battersea until the sport was abolished in 1910, mainly due to the humanitarian insistence of Queen Alexandra.

The popularity of pigeon-shooting and game is reflected in the orders for cartridges. A. G. Spalding of Chicago and Spalding Haywood Arms of Colorado, in the USA, and the Union Espanola de Explosivas, in Spain, took over 50,000 cartridges annually, loaded with different amounts of EC and Schultze powders, between 1898 and 1905; 99,000 cartridges being hand-loaded for this latter agency in 1903. The consignments included 6,000 brass 12-bore cartridges loaded with EC powder for the King of Spain and about half that number of 16-bore cartridges loaded with Schultze powder for the Infanta Isabel. The committee of the Pigeon Club of Buenos Aires were supplied with 5,000 cartridges of No 6-7 shot in the same year. There was also a customer in Lisbon — a J. M. de Fonseca — with whom, as James was very partial to his glass of port, a useful relationship was established.

In 1906, however, the reputation of the Purdeys was suddenly challenged by an unexpected and most unwelcome lawsuit when, on Saturday 15 December, Mr John Robertson of Boss & Company took James Purdey and Sons to law in the Chancery Division of the High Court before Mr Justice Parker, claiming an injunction restraining Purdey's from infringing the patent taken out by Boss in 1894 relating to the invention of a certain design of single trigger. The case lasted until 26 February the following year, Purdey's being represented by Mr A. J. Walker, KC, and Boss by Mr T. Terrell, KC. There was nothing new in the single trigger itself, as flintlock guns had been fitted with such devices many years before this date, but since it was found that both barrels tended to go off together few guns were made in this way. Guns with single triggers had always been a bugbear to the Purdeys; such devices are just one more thing to go wrong, and they always believed in the simplest of mechanisms. Not only that, the single trigger can be adjusted to one individual's style and reactions, but the same mechanism may not necessarily suit another customer who buys that gun secondhand. Yet although more trouble had been caused by failure of the single-trigger mechanism than any other cause, Athol felt that, as other firms had one, Purdey's should have one too. One of the actioners, William Nobbs, had invented a mechanism in 1883 which was a failure, and the 'Old Man' would have nothing to do with it. A second attempt in 1894 was a success and was patented in that year. Suffice it to say, that Athol Purdey suspected Messrs. Boss of contravening this patent for a 'three pull' single trigger and went to see Mr Robertson in March 1897.

Robertson had, as already stated, also taken out his patent in 1894 and, according to his evidence at the trial, Athol was at that meeting insolent in his approach saying, 'Look here, Robertson, I hear you are saying that you are the inventor of the three pull mechanism and I won't have it.' Robertson then replied: 'It is true I worked for your father for many years and I don't think his trade suffered at my hands. If you can't speak more civilly you can go out. There's the door.' Athol then calmed down and showed him the specifications of Nobbs' invention, adding that Purdey's, as a firm, did not think that single triggers had come to stay but they were doing something to show that they were alive in the matter. Robertson said that he did not agree that the specifications shown to him were the same as his own, and closed the interview by saying to Athol: 'Do you know what I think you have been doing? You have been laying ostrich eggs and have only got sparrows.'

Nothing further happened until Boss brought the case in 1906. Many witnesses were called, and elaborate working models of actions and single-trigger actions were examined by the bench. Athol Purdey repeated his remark, in the witness box, that he did not believe in single-trigger guns, the sale of which, he thought, was decreasing. He explained the modifications he had made to his guns and denied that they were any imitation or encroachment on the plaintiff's specification. The *Shooting Times and British Sportsman* and *Arms and Explosives* gave full cover to this splendid row between two leading gunmakers and in the end, on 26 February 1907, Mr Justice Parker gave judgement in favour of Purdey's, clearing their name and awarding them costs. Nevertheless, it would have been much better if it had never happened as the gun trade had taken sides and ill will, thus created, takes a long time to die down. Quite apart from that, a case of this nature which implied dishonesty and double dealing did no good to the name of Purdey until the matter came to a satisfactory conclusion.

But 1906 also brought more profitable business for in that year the new Prince of Wales, together with the Princess, paid a visit to India. The book entry is headed 'The Indian Outfit', and the Prince took with him his own three 12-bores, one of which was fitted with new barrels choked for ball cartridges, presumably for pig shooting; a new double gun with choke barrels; and a .450 hammerless Express cordite double rifle, together with cases, cartridge bags, covers, oil, tow and all the usual accessories. The names of the Indian princes who ordered guns in that year read like a durbar: Sir Aga Sultan, Muhammad Shah, Aga Khan; Mir Sir Faiz Mohamed Khan; the Maharajah of Kuch Behar; the Gaekwar of Baroda; the Maharajah of Alwar, Sahib Bahadur; the Maharoa of Katah; the Maharana of Oadeypore; the Mir of Khairpur; the Maharajah and State Council of Jammu and Kashmir. Mustafa-al-Mamelek of Teheran also ordered, as did His Highness Ras Makunan of Ethiopia.

Back in England, game shooting was going from strength to strength. Individual orders for cartridges of 10,000 per season are commonplace in the books of the time, and the cartridge-loading shop was busy far into the night. The orders for guns never slackened and the profits boomed, but now James the Younger, the great 'Governor' was continuing to lose his grip. He had gone down to Margate and was being looked after by his old servant Alice Wells. On 19 March 1906 she wrote:

Dear Athol

I gave your message to Mr Purdey and he asked me to thank you. When I said should I give him your love [sic] he said 'I don't think I *can* send my love to a Man but you do what you think best.' He does say funny things; when I told him at Breakfast that it was his birthday and he was 78, he said 'Keep that knowledge to yourself.'

His mind was really going now and he kept forgetting names but always

Lord Savile of Rufford Hall, Nottinghamshire; one of the great shots of the Edwardian era, and a very good customer of Purdey's

perked up when he went out in his carriage. At the time Athol was acting as a juryman, so could not attend his father's birthday, but there were two more birthdays still to come before Athol assumed the complete control of the firm.

During May and June 1907, Athol was very busy advising on the reorganisation of the gunroom layout at Windsor Castle. Having been asked by Lord Farquhar for his views, he came up with the suggestion that good serviceable gun cabinets in light oak should be constructed, one to hold six guns for the King, a similar case for the Prince of Wales, and a third to hold six more guns for the use of any royal visitors. Another general gunroom should be provided for other guests, with accommodation for eighteen guns, and strong shelves to hold cartridges etc. Special care was to be taken to allow for the differences in stock lengths for royal visitors because of the long stocks on the guns used by the King of Norway. Lord Farquhar wrote on 23 July 1907, saying that these plans had been approved by the King and would Mr Purdey see that all the work was completed in time for the next shooting season. At the same time, special gun-racks were fitted by the Daimler motor company to the King's shooting brake.

King Edward VII out shooting

James the Younger had drawn up his last will on 14 March 1903. He had made a will on 6 August 1879 in which he had included his eldest son James, made provision for Athol and Cecil and also for the firm in the event of James's death; but it was not until 1903 that he revised his bequests. He appointed his widow Julia, Athol and his son-in-law, Walter Green, as executors and trustees. His first wish was that he should be buried in the family grave in Paddington Cemetery at Willesden and that his funeral should be as simple and inexpensive as possible. After leaving all his personal jewellery, pictures and possessions to Julia, along with her own paraphernalia and clothing, he included the leasehold of 28 Devonshire Place together with his carriages, carriage horses and stabling. She was also to receive £500 at his death, £1,750 a year during her widowhood and £400 a year if she remarried.

He then bequeathed his business of gunmaker, including his leasehold factory in North Row, his bank debts and the money standing at the banker's to the credit of the business, to Athol and Cecil in proportion of two-thirds and one-third respectively; but not including Audley House unless Athol should take an underlease from the trustees at a clear yearly rent of £700 for a forty-two year period from his decease.

He gave James's daughter Dorothy Irene (his grand-daughter) an annuity of £100 for maintenance and education, and Percy got the income of £150 from a special trust fund, exclusive to himself. His daughter, Florence Green, who being childless had adopted a daughter, Ashdown Green, was given £750 and allowances for her child; but Constance had already received money from him and so got no special mention. Lionel Bateson Purdey, aged seventeen when the will was drawn up, was to receive £750 as soon as he should reach the age of twenty-one.

The whole of his properties, apart from the gunmaking side of his affairs were, therefore, made into a trust for the benefit of his sons and daughters, with the explicit exclusion of poor Percy, the sons receiving double the shares of the daughters. Money requiring investment for this trust could be invested in the usual government stocks and specifically in any company established in England for the purpose of buying land in England, Wales or 'the Australian Colonies'. He had always had the greatest faith in the financial expansion of the British Empire, and included the Indian railways as another acceptable investment. Should Athol or Cecil wish to take any interest in the residue of his property they were to bring into account £60,000 and £30,000 respectively.

On 19 March 1907 the family gathered at Devonshire Place to celebrate the seventy-ninth birthday of the Governor. Sefton Purdey had drawn a folio of sketches which were given to Julia and the 'Old Man', two of which are reproduced in this book; but this was the last time he was able to cope with a large gathering. Two years later, on 13 March 1909, James Purdey the Younger died suddenly after a very short illness, in his eighty-first year. The obituaries that appeared in the *Field*, the *Country Gentleman*, and *The Morning Post* traced his career from his earliest years to the position he held at the top of his trade. They mentioned the respect in which he was held by his workmen, rivals in the trade and customers alike, these sentiments being confirmed by the letters of sympathy Athol received from many old friends and customers. The Prince of Wales sent a telegram to Athol signed by Derek Keppel, Equerry in Waiting, offering his sincere sympathy to the Purdey family at the loss of their father. Sir Ralph Payne-Gallwey wrote:

> I consider his absolute integrity, his invariable courtesy, and his expert knowledge of his business — far the most famous and justly respected one of any of its kind in the world — made him stand out as a fine example of what the head of a great firm should be.

Mr Remington-Wilson added: 'To be absolutely first in one's line is no easy thing and he was easily that.' Lord Ancaster, Lord Ashburton and Walter Winans wrote charming letters in the same vein, while young Tom from West Downs wrote to his father:

> My dear Daddy,
> I was very cut up about your telephone message. I did not know Granpar was even ill. But of course it must have been a great relif [sic] that the poor old Gentleman was off your and your relations mind. For my Birthday I want lots of small passels [sic] you know what I mean any little novelty you see. See you Saturday lunch at George. Your loving son. Donald.

Lord Bath wrote:

> Lord Bath begs to thank Messrs. Purdey for their cheque, also for their letters for which he is much obliged. Lord Bath would like to take this opportunity of saying that it was with great regret that he saw the announcement of Mr Purdey's death.

Nevertheless, the shooting public and the gun trade were astonished when the details of his fortune were published. He had always been accepted as a rich man but the details of his will, when released, took them by surprise. In May 1909 *The Globe* had a heading 'Will of a Rich Gunmaker', and *The Times* 'Gunmaker's Fortune'. The *Westminster Gazette*, publishing 'Today's Wills', on 25 May says:

Mr James Purdey, Eighty, of South Audley Street and Devonshire Place W. late head of the firm of Messrs. James Purdey and Sons, Gun and Rifle Makers, an original member of the Constitutional Club, and one of the oldest members of the Royal Yacht Club, left estate valued at £200,289 gross, of which £178,248 is net personally.

The Very Rev. Herbert Mortimer Lubbock, D. D., Dean of Lichfield, who died on March 24th, left £24,464.

This was a very great deal of money at 1909 values.

Thus ended this remarkable man who was born in the era of flintlocks, learnt his trade in the time of percussion guns, moved on through the first 'breakdown' guns, understood pin-fire guns, needle-guns, centre-fire guns and invented the bolt and top lever which made all these new things safer; who realised the benefits of the Beesley action, grasped the niceties of the ejectors, encouraged then rejected the concept of the single trigger and did all he could for the safety of the public through the Gunmakers' Company. And not only his own but his family's reputation was acknowledged in the *Field* of 20 March 1909 in which, at the end of the eulogy on his career, the obituary concludes with:

> Mr Purdey was an Artist in his subtle appreciation of all these things, but nevertheless the desire to pay due tribute to the dead must not permit an injustice to be done to the living. Just, in fact, as Mr James Purdey proved himself the able successor of his father, so those who are now carrying on the business may be regarded as capable exponents of the family tradition of gunmaking.

He was buried on Saturday 20 March 1909 in the family vault in the old cemetery, Paddington (vault 1.H No 756 by modern numbering) which contained the remains of his father and mother, his first wife Caroline, two sons by his first marriage, and one son and three daughters by his second marriage. His second wife Julia was to join them two years later in January 1911, and after that the vault was sealed. It is now covered by two flat stone slabs. Originally it was surmounted by an ornate marble tomb elaborately carved with pediments and angels, but in 1970 the firm was informed that the fabric of the tomb was in a very bad state of repair and needed extensive restoration work . So high was the quotation for carrying out this work that it was considered better to remove the marble which had become a source of danger to the public. Unfortunately there is no photographic record of its appearance.

Athol was now truly the head of the business and of the family. Ably helped by Cecil, the customers had full confidence in his abilities and there were no doubts that he would be able to maintain the quality and style of the firm. In the same year, in fact, the King of Sweden ordered a pair of guns, and in 1910 he granted his warrant to Purdey's. Also in 1910, Sir Basil Brooke, the future Lord Brookeborough and Prime Minister of Ulster, grandson of the famous big-game shot, Sir Victor Brooke, became a customer at the age of twenty. Lord Durham ordered a pair of 12-bore guns for his friend Mr Stobart and a set of three for himself, together with a .400 rifle, and another pair of 12-bores for his son; and the Queen of Spain ordered a pair of guns for Prince Maurice of Battenberg.

In May 1910 King Edward VII died at Buckingham Palace. His great popularity was manifest by the remarkable crowds at his funeral procession through the streets of London, and the touching tributes from people throughout his Empire. The photograph opposite shows the new monarch, King George V surrounded by his fellow sovereigns at Windsor Castle on the day of the funeral, 30 May 1910. Within a few years, the whole way of life and government as represented by these nine kings was to be destroyed, but James Purdey and his son Athol had built guns for each of the rulers in that photograph, Kaiser Wilhelm II included, and Athol was very proud of this fact.

King Edward had been particularly loved by the Purdey family for his amiability, courtesy and consistent loyalty to his gunmakers, and Athol was very upset at his death. However, his relationship with the new King was just as close and his personal admiration for him unbounded — in 1910, George V ordered a 16-bore gun for Prince Albert, the future George VI. No one at that time, however, could dream of the changes that were to take place to the established way of life of England or the fortunes of the firm during the course of the new reign, and indeed in the remainder of Athol's life until his death in 1939.

The Nine Kings, photographed in 1910 at Windsor after the funeral of King Edward VII. Purdey's had built guns for all of them. From left to right, standing: King Hakkon VII of Norway; King Ferdinand of Bulgaria; King Manuel II of Portugal; Emperor William II of Germany; King George I of Greece; King Albert of Belgium. From left to right, sitting: King Alfonso XIII of Spain; King George V of England; King Frederick VIII of Denmark.

WORLD WAR 1914–18

The Coronation of King George V and Queen Mary was to take place on 22 June 1911, and decorations and displays for the great day went ahead at Purdey's, the centre of the crystal display being changed from 'E VII R' to 'G V R'. Athol was thrilled to receive a letter from the Keeper of the Privy Purse in the same month of June forwarding a medal to be worn in remembrance of the royal event.

By this time, young Tom was at Winchester though his getting there had not been easy, having failed the Common Entrance examination. However Winchester had allowed him to join Mr Fort's 'B' House in May 1910 at the age of thirteen years and one month. He was very homesick and found that those still there who had known his brother James were only too glad to redress any past wrongs on him. James was good looking, good at games and rather dashing; Tom was small, fat, spoilt and very self-opinionated. It is understandable therefore that when he said he saw no good reason 'why he should learn inane words for school things', that he should have been beaten by the Prefects well and often, to quote his Diary. However, on his first leave from school he had gone up to London for King Edward's funeral, viewing the procession from St James's Palace. Back at Winchester his friends who supported him were Guy Dodgson, who was later to be killed in World War I; Jock Davidson, later Chief Constable of Kent but in those days known by his contemporaries as 'Bum-fluff'; a couple of Martindales; Charles Miller; and Justin McKenna who had been at West Downs and 'being very popular with everyone often helped me — who certainly was not'. Later that summer, Tom had caught measles and scarlet fever while on holiday at North Berwick, and had to stay there until November with Mabel; but this further complicated his life at Winchester as he missed the short half. Nevertheless, some excitement came his way, as he records that Dr Crippen was caught and hanged during his convalescence. And during April 1911 the family moved into 7 Clifton Place, Sussex Square, which was much better, in Tom's opinion, than the old house at 20 Queensborough Terrace.

Jim had joined the 21st Lancers from Sandhurst on 25 March 1911, and

Jim in 1911, in the uniform of the 21st Lancers

sailed for Egypt almost at once in the troopship *Soudan* to join the regiment from Southampton; Tom, Athol and Mabel seeing him off from the quayside.

Athol was ill during April so Mabel took her husband and Tom to Biarritz, staying in Paris on the way out, and Tom found the one real relaxation of his life when he discovered, at the age of fourteen, the joys of fishing in the Lac de Chambertin. On his return to Winchester, where Mr Aris had by now taken over as housemaster, he writes that 'perhaps the greatest

York Cottage,
Sandringham,
Norfolk.

December 23rd 1913.

Dear Mrs Purdey,

I thank you so much for sending me the pretty little water-colour of Eton. I like it very much and it is _so_ pretty.

> I am very pleased to have my brothers home for Xmas.
>
> With best wishes
>
> Believe me
> yours sincerely
>
> Mary

A letter sent from Princess Mary to Mabel in 1913

event in my young life, was that I started to fish for trout on the Itchen'. On his return to the college everything was even more difficult. The short half was awful and he was desperately unhappy: 'Everything worse and worse — could do nothing right — crisis over work.'

Fortunately for him, Mabel took notice of his plight and approached Eton College with the result that the headmaster, Dr E. Lyttelton, wrote asking her to bring Tom down to the school and he would set him some papers to complete, adding that 'It may be possible to admit him owing to the fact that we have arrangements here for handicrafts work in place of some of the book work, and I understand that that is likely to be suitable for him.' Tom passed the tests, and on 20 December 1911 his future tutor, A. M. Goodhart Esq, wrote to Mabel from Godolphin House:

> I can take him *next* half and hope that I may be able to keep him altogether. You will understand that some boys are coming September next year who have been down for ten years or more and that I must not disappoint them. But I hope that if boys leave there may be room for both. It is not always a good thing for a boy to change schools but I have had one case of a Charterhouse boy who came on to Eton and has done very well.

In January 1912, therefore, at the age of fourteen years and ten months, Tom entered his new school and later wrote: 'I can never be grateful enough to Mother for insisting on Eton.' He had only managed to take Lower Fourth, the lowest form but one in the school, after two years at Winchester, and so was at least two years older than anyone else in this division, with the result that he did rather well. He was very happy in his house, and he put this down to the personality of his 'Dame', Lady Georgina Legge, a sister of Lord Dartmouth, who was greatly loved by everyone who knew her. Tom won the House Fives, learnt the piano and was given his House Colours for the Field Game. Jim came back on leave and taught him how to drive a Renault on Romney Marshes during the holidays. He also took riding lessons at Smith's Riding School, but found that he did not care for horses and they did not care for him. The year 1913 opened well when he won the Silver Bugle competition playing 'Hello Ragtime', and 'Gypsy Love', and Princess Mary had tea with him in his room. He attended his first Gunmakers' Company dinner on 3 April, and when on 16 June the King and Queen paid a visit to Eton he had tea afterwards with Princess Mary at Frogmore. He was a 'wet bob' and greatly enjoyed rowing on the river where he surreptitiously had his first smoke. And so the last wonderful year before the war went by with Tom enjoying life to the full. In fact so much did he enjoy life, getting his House Colours and his 'Boats' for rowing, that he did very little work, and became

James the Founder's work tools

One of the set of three miniature guns built by Harry Lawrence in the Silver Jubilee year of 1935. This gun is exactly one sixth the size of King George V's 12-bore hammer gun, and all three guns work and fire cartridges specially made for them.

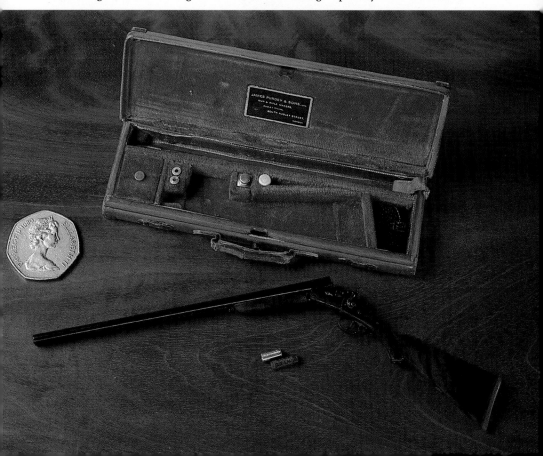

Detail of a Purdey 12-bore pinfire double-barrelled game gun, no 6829, built in 1864. Weight 6lb 14oz; barrel length 30in

The engraving on one of the pair of 12-bore guns given to Lord Snowdon by The Queen on his marriage to Princess Margaret in 1960. The engraving was designed by Carl Toms, and executed by Ken Hunt

odiously spoilt with the result that he was sent for by the headmaster and, on his seventeenth birthday, given twelve of the best by the captain of the House.

The summer half of 1914 was glorious. He was elected to the House Library and rowed in the Procession of Boats. He saw Gustave Hamel loop the loop over the King at Windsor, and went to Lord's where Eton beat Winchester by four wickets. Holidays at a house near Dartmouth were great fun, but on 4 August Britain declared war on Germany and, on returning to Eton on 17 September he found everyone miserable and unsettled. At seventeen Tom decided to try for Sandhurst, took the examination, passed, and on 24 October left for the army, proudly wearing his Old Etonian tie.

At Sandhurst he trained hard until, on 17 February 1915, he was commissioned as a second lieutenant in the Argyll and Sutherland Highlanders at the age of seventeen. Complete with valise, kilt, claymore, glengarry and white spats with tartan stocks, he drove to Purdey's to show off to his Uncle Cecil and then joined the third battalion of his regiment at Abbey Wood near Woolwich, where he was greeted by his colonel, Woodridge Gordon. He was soon taken down several pegs as no one spoke to him in the Mess, and he was drilled and disciplined by Sergeant Major Bodie. On 22 March, Tom was warned that he was to take a draft of soldiers to France, and on the morning of 21 May, he left with his draft of the Gloucester

Tom saying goodbye to his Uncle Cecil before leaving for France in 1915

Regiment for Southampton. Athol and Uncle Cecil went down to see him off and gave him lunch and lots of port at the Dolphin Hotel; and then, hung with revolvers, compasses, periscopes, wire-cutters, a torch, but no claymore, 'very tearful and horribly frightened', he said goodbye to his relations at the dock gate and sailed for France.

After a short time at Rouen Tom joined the 91st Highlanders at Boozebaum on 22 May and, with Colonel Kirk in command, marched for the Front Line. One lovely evening with a clear blue sky he got his first sight of war — Ypres burning very red before him. The shelling was terrifying and they endured a gas attack which was 'most unpleasant'. The regiment moved into trenches at Chapelle D'Armentiers with the Hun only about eighty yards away. The latter put up a sheet with a large notice, 'Who won the Battle of Waterloo?', and the Highlanders shot every letter away before 6.30am.

Tom's eyesight had deteriorated badly when he was at Eton, and he had worn spectacles from the age of fifteen. This disability was now a real handicap, as he could not see details clearly at any great distance. His colonel and adjutant were just as concerned and, after a very sad interview, a protesting Tom was sent back to Rouen by ambulance where Sir William Lister told him he was unfit for active service. He developed headaches and heart trouble and was sent to Osborne House, in the Isle of Wight, for treatment. After a short spell with the 13th Scottish Rifles (the Bantams), he was posted to Stirling Castle, the depot of his regiment, on 20 January 1916, as adjutant to Colonel Sir Philip Trotter. He was not quite nineteen.

While Tom Purdey had been going through these extraordinary adventures, the firm itself had had to face a complete change in its affairs. Profits and orders for new guns both from home and abroad had continued as usual during 1912 and 1913, but after that date things altered with alarming suddenness. The net profit for 1912 was £8,024 and in 1913 it was £8,216. But in 1914 it fell to £4,844 and in 1915 to £725, before war work could be organised to save the firm from bankruptcy.

Until the very start of the war, business as a whole went on much as normal. The usual numbers of cartridges were loaded and supplied; shoots took place and guns were repaired and maintained. There was light relief when one of the firm's debtors, Count von Wolff-Metternich, nephew of the German Ambassador in London, stood up during his trial in Berlin and told the judge in his defence that he had contracted his debts in certain expectation of a rich marriage to the heiress of a fortune of £250,000. Athol wrote off the amount the young man owed him as a bad debt.

When war was declared in August 1914, everything changed completely. Up to that moment, Athol had relied on a stable national economy from which rich customers with opportunities for sport would buy his guns;

suddenly, the whole basis of his trade vanished. His work force, then numbering about ninety craftsmen and dependent on a constant supply of orders, found these orders had ceased overnight. A third of his best men promptly left the firm to join the colours, twenty-five trained men between the ages of twenty and forty-five joining the Royal Army Ordnance Corps and becoming sergeant artificers. The future looked desperate.

All orders from the German Empire, the Austro-Hungarian Empire and their allies of course ceased at once, and these countries had constituted a large percentage of the firm's business. The Russian Empire, fighting on the side of the Allies, could not be reached due to the difficulties of communication across war-torn Europe although, strangely enough, Finland continued to place orders until 1916. Russian gun orders that were building when war was declared were completed and delivered via Bergen in Norway, Count Paul Schuraloff's gun being delivered to his house in St Petersburg in 1915, and the Moscow Sporting Company of the Société des Chasseurs taking delivery of three guns and a light .400 bore rifle during the same year. The Moscow firm of E. Bernhard and Co, who had for many years enjoyed a discount on orders of 20 per cent, ordered six to eight guns per year until 1915, the final gun, No 21465, being despatched to them on 15 January 1917. No payment was received, and the debt was eventually written off in 1924. From 1917 onwards all contact with Russia ceased.

Italians had been buyers of pigeon guns since 1900, and a good trade had been built up in game guns from 1909 onwards, but the war stopped all orders. Wolfredo Boldi of Florence for instance, who had taken an average of six guns a year, cut their requirements to two in 1915 and then ceased to order any more. Portugal, a major customer, ordered nothing, and the orders from India dried up. From France the story was the same. Almost every private account closed in 1914 and only a few were to reopen in 1920. Guinard and Cie of Paris, for years close friends and agents to the Purdey family, who had ordered on average ten guns a year from 1910, ordered one gun in 1915 and then no more for the duration of the war. The growing market in America shut down completely after delivery of five guns on order in 1914 for the New York firm of Von Lengenke and Detmond.

In England, naturally, the position was the same; there was a cutting down of all unnecessary expenditure. And although shooting continued in a very small way during the duration of the war, it was, with very few exceptions, rough shooting that was enjoyed. There was, however, one near disaster to Purdeys' English customer relationships. The Duke of Buccleuch came during 1916 to have his measurements checked with the 'try gun'. Athol saw him personally in the Long Room and asked him to point the 'try gun' at his eye and to 'just step backwards a little'. The Duke did so and vanished, a loud crash denoting that he had landed in the Gun

Room at the foot of the staircase he had failed to notice behind him. As he was one of the best customers Athol was appalled. However the Duke was picked up and dusted down none the worse for wear and, although he had previously been 'health conscious', never referred to his complaints again.

Customers who had been used to shooting off 20,000 to 25,000 cartridges each year cut their orders by a quarter in 1915, and then to 2,000, and this for pressing reasons. Park land was being ploughed up for food and coverts were felled for much needed timber. Pheasant rearing virtually ceased as keepers, beaters, loaders and the sportsmen themselves, joined their regiments. Taxes were increased to pay for the war, and there was no heart for expensive amusements like shooting while men were dying in France. Lord Ripon's account for example, which normally stood at over £200 per year, fell to a mere £18 17s in 1915, and £23 14s 6d in 1916. Lord Dalmeny, another excellent customer, was serving in the Middle East on General Allenby's staff, and his account which had averaged £170 a year between 1910 and 1914, fell to 4 guineas in 1916; the Duke of Richmond spent £1 13s 6d in 1915 and only 8s 6d in 1916. Taken over all, though, the English accounts fell to about a quarter of their prewar level during the war years. Not all fun ceased however, as is shown by the record grouse bags at Abbeystead on 12 August 1915 when nine guns — including Lord Sefton, Sir Harry Stonor and my great uncle, Edward Beaumont — shot 2,229 birds by 2.30pm, at which point the party ran out of cartridges, Sir Harry having shot 457 grouse to his own gun. This remarkable day's sport took place on Littledale and Abbeystead in Lancashire, and the two following days on Mallow Dale and Tambrook the same guns bagged 1,763 and 1,279 grouse respectively.

With a desperate financial stalemate facing him, Athol knew that not only must alternative work be found for his men, but at the same time he must do something to help the war effort. He approached the military authorities in February 1915, and in April the Secretary of State for War placed an order with Purdey's for the manufacture of muzzle-protectors for rifles. No one had envisaged the complications of trench warfare, so the likelihood of rifle barrels becoming full of mud from being jammed into the sides of the trenches had naturally not occurred to the armourers. Muzzle-protectors, consisting of a spring-activated flap which closed over the end of the bore of the rifle, were designed to obviate this hazard. Harry Lawrence, son of the factory manager, and who had started his apprenticeship at the factory in 1914, was transferred to the basement of 84 Mount Street where the machinery for this work had been installed, and 18,000 were produced in April 1915 alone. As the men in the factory were so skilled, it was not necessary to give them long retraining for this type of work, and so when the next order was received, which was to fit telescopes to snipers' rifles, they felt quite at home. No extra men were taken on, although apprentices were still taught by the older men who stayed behind.

Knowing Purdeys' reputation as rifle makers, the War Office had asked Athol to fit, test and deliver these telescopic sights as soon as possible. The first batch of six were, ironically enough, of German manufacture and were completed and delivered to the War Office on 28 June 1915; whereupon a further order was issued for the fitting of English telescopes to service rifles and 499 were mounted, tested at the grounds, and supplied with leather cases between July and December the same year. At the same time, the muzzle-protectors continued to be produced and 94,000 were completed by the middle of September.

The main credit for the successful reorganisation of the factory staff to deal with this sudden flow of completely different work, must go to the new factory manager, Ernest Lawrence. He had to ensure that the standard of the work supplied to the War Department was of the highest quality and when, later in the war, precision sights were made to the order of the Royal Flying Corps, his own genius of invention and adaptation was to prove invaluable. At the same time he had to complete the building of guns still on order and any new gun orders as they came through, and also to continue to train gunmaking apprentices for the future in spite of the fact that the boys left to join up when they reached the age of eighteen, just as they were beginning to know their jobs. All of this he achieved with great skill and patience so that Athol was able to take further war orders, and between January and July 1916 some 944 telescopic rifles were completed. At this point, the requirements of the Ministry of Munitions began to take precedence, and the machining and manufacture of magazines for the Mark 2 Lewis gun kept some of the men very busy for the remainder of 1916, while others were employed on the manufacture of the Hutton sight for Lewis guns. On top of that, 'spade-grip' thumb-triggers with safety catches for Lewis guns were produced at the rate of 200 per month, and special loading devices were made for the same weapon.

Fortunately for the gun side of the business, there was still one country which, being neutral, continued to place orders — namely, Spain. Eduardo Schilling ordered eleven new guns including sets of three, in 1915, and twelve more in 1916. Casa Paolo of Madrid and Emeterio Echererria of Eibar ordered fourteen between them in 1916, one heavily chased for £150, and individual sportsmen such as Conde Romanones and the Marques Fuentes el Sol had guns built during 1917 and 1918. In the same period King Alphonso, 'one of the best friends Purdey's ever had' as Tom wrote in the ledgers, continued to order his cartridges, taking 40,000 between 1915 and 1917 and 100,000 in 12- and 16-bores in 1918. Somehow Athol and Lawrence between them managed to build all these guns and load these cartridges within the delivery times expected, and also to keep up war production to the standards and delivery demanded by the War Office.

King Alphonso XIII – 'the best friend Purdey's ever had'

While his father and his Uncle Cecil were fighting to keep the firm solvent, Tom's service career had taken a new direction. His life as adjutant to his regiment was extremely enjoyable and Sir Philip Trotter, his commanding officer, was charming to him. Mabel went up to Stirling and took a small house in Victoria Place to be near her son, and it was at Stirling that Tom first felt the excitement of flying, for Major Sholto Douglas of the Royal Field Artillery arrived there in the spring of 1916 to form 43 Squadron, Royal Flying Corps. Tom was taken up for his first flight by

Lieutenant Lees, known as 'Uncle', later Air Officer Commanding 2 Group during World War II, and never looked back from that moment on.

After Mabel's return to London Tom was boarded in Glasgow; and although unfit for active service was getting desperate to return to some form of action. Sir Philip Trotter, at the age of seventy, was flying daily, and Tom made up his mind to join 43 Squadron as 'Duggie's' adjutant. He applied and was seconded on 10 November 1916 as adjutant to 43 Squadron, RFC at Netheravon in Wiltshire — a big fighter station equipped with Bristol Scouts. Sholto Douglas was in command, all the men were new and all the aircraft old. It was here that he made the closest friendship of his life when Captain Harold Balfour, King's Royal Rifle Corps, flew in to take over as flight commander. Having left Blundell's School on his seventeenth birthday, 1 November 1914, Harold had been gazetted as a second lieutenant in the Special Reserve of the King's Royal Rifle Corps in the spring of 1915 and went out to France into the front line on 4 August. In 1916 he transferred to the RFC 29 Squadron, and fought at Arras before being posted to 43 Squadron at the age of nineteen.

Jim Purdey, meanwhile, arrived back from India in December 1916, very ill and covered with boils, and was sent immediately to the Great Central Hospital. Tom writes in his Diary: 'So glad to see him — almost a stranger.' The brothers had not seen each other for six years so it was hardly surprising that they had very little in common.

Jim's life during those years had been varied and very interesting. As early as the spring of 1915 there had been signs of unrest on the North-west Frontier of India as the Germans and Turks incited the Mohmand tribe of Afghan origin to advance through the Khyber Pass to attack the Punjab. The 21st Lancers was the only British cavalry regiment left in India and when matters came to a head in August 1915 Jim's Squadron 'B' marched out north from Risalpur, in appalling heat, to join the field force ordered to repel these attacks. On the morning of Sunday, 5 September, the regiment deployed at the Shahkador crossroads. The enemy were entrenched on the other side of a canal, so Jim was instructed to take two troops, cross the canal and sweep round a small village on the bank. Colonel Scriven, with the rest of the regiment, attacked the enemy from the front and charged over the canal. The water proved to be deeper than anticipated and considerable difficulty was experienced in crossing, the tribesmen engaging our men in close combat with knives and swords as they scrambled up the bank. The Lancers were outnumbered by at least five to one, but nevertheless forced the crossing and drove the enemy back into the crops of maize, very thick and standing eight feet high, in which terrible fighting ensued.

In the meantime Jim, leading his troops through dust at least a foot deep which produced a cloud like a London fog, temporarily lost his way and, when he did find the canal, his young mare refused to either jump it or wade it. They eventually forced her across only to find that the further side had been recently irrigated and the horses sank to their saddle girths in the mud; all this under heavy enemy fire. The delay meant that Jim and his men were too late to play a significant part in the action.

The main attack was completely successful in driving the enemy back, but there were many British casualties including Colonel Scriven who was shot through the heart by a tribesman after being rescued from under his fallen horse by two of his Lancers. Private (Shoeing Smith) Hull was awarded the Victoria Cross for his gallantry in saving the life of Captain Learoyd, the adjutant, whose horse had been shot under him. The enemy clearly having had enough, the regiment was withdrawn to Risalpur again on 25 October.

On the return to the main depot, Jim was appointed ADC to the Governor of Bengal, Lord Carmichael, and in September 1915 he travelled to Calcutta and took up this appointment. He was a great success and thoroughly enjoyed his duties, but felt frustrated while his contemporaries were on active service on the Western Front, and he spent some time trying to get transferred to a regiment that was fighting in Europe.

During 1916 his health deteriorated and he returned to England for a period of convalescence in the Great Central Hospital. He found that there was a vacancy for a captain in the 12th Royal Lancers, applied, was accepted, and went out to join his new regiment in the trenches in France at the end of June 1916. He was invalided home in April 1917, when his first marriage took place, to Harriet Patricia Kinloch, only child of Major Henry Kinloch of Gilmerton, Midlothian. He then served at the War Office until the end of hostilities.

Tom had just gone off again to war. No 43 Squadron was ordered to France and on 18 January 1917 the aircraft under the command of Major Sholto Douglas flew across the Channel, while Tom and Harold Balfour took the ground crews and transport by rail and sea to Le Havre. They all met up at an airfield at Trezennes. The cold was awful and the accommodation worse. No 40 Squadron, *in situ* under Robert Loraine, helped them as much as possible, but their planes were old-fashioned Sopwith two-seaters and their guns froze solid in the frost; opposing them were new German Albatross fighters. The General Officer in Command was 'Boom' Trenchard, who settled them good and proper, for on a visit of inspection he said: '43 Squadron, stop grousing or go home!' As Tom comments in his Diary, the squadron did neither but started fighting, and after their first 'show' during which they suffered casualties but shot down some German planes, 'Boom' came down in person to congratulate them.

Harold Balfour shared a hut with Tom who was superb in the way in which he tried to keep up the morale of the pilots in the dreadful spring of

Tom and Harold Balfour in France in 1917

1917 — playing the piano, singing, drawing amusing cartoons and on occasions allowing them to hold him by the ankles and shake him upside down in his kilt! The cold remained intense and casualties were appalling, thirty-five pilots being wounded or killed during March and April. Up against Richtofen and Goering, in superior aircraft, doubts grew as to the reliability and performance of the British aeroplanes. Many of the pilots had the idea that the Sopwith two-seater had a basic structural weakness and could not be put through violent evasive manoeuvres. Sholto Douglas was determined to show otherwise, so on Tom's twentieth birthday, 22 March 1917, he took him up as his passenger in front of the whole squadron, and looped the loop no fewer than twelve times in succession to prove his point. As Harold Balfour, who later won the MC, said, if he had not been in the hands of an expert Tom would probably not have survived to run the firm in later years! But survive he did, and was delighted when, in the following month, Jim married Patricia Kinloch.

During these stirring times for Tom and Jim, Athol Purdey was suddenly sent for by the War Ministry in the autumn of 1916 and asked to visit Professor Norman at a certain room in the Cecil Hotel in the Strand. As it was to be on a technical issue, Athol took Ernest Lawrence as his adviser. Professor Norman had been hard at work at Cambridge University developing a special sight to be fitted to our fighter aircraft, and Athol was asked if the Purdeys' craftsmen would be able to build it and the windvanes which were part of its design. Brackets to hold it in place would also have to be provided, and an illuminated ring with electrical accessories for night fighting would be installed. This was just the sort of work best suited to the ingenuity of Lawrence and he set to work at once to start production. So successful were the initial trials that the first completed sight was fitted to the gun on the plane flown by Captain Leefe Robinson when he shot down the first Zeppelin on the night of 3 September 1916 at Cuffleys, near Barnet, Hertfordshire. Over 4,500 of these sights, with their attendant brackets, were made for the Royal Flying Corps before the war ended in 1918, the work force of the factory being redirected to this new form of manufacture. As a result of this success other forms of sights were ordered such as the Neame, the Neame Illuminated, the Hazzleton muzzle attachment, and the Cox's cocking handle for the Vicker's gun, as well as the Hutton night-flying sight in August 1917, and a variety of brackets for Lewis gun attachments to aircraft. All these articles were made in their hundreds, both in the factory in Praed Street and in the basements of South Audley Street and Mount Street. And in 1917, in order to speed up production, two lathes were installed in what is now the accessory shop in 84 Mount Street and, for the first time in the history of the firm, five women were employed under the supervision of William Clarke to work the machines.

For such a small business, Purdey's had certainly acquitted itself well in the war effort, and though by no means war profiteering, had managed to build up its financial position so that it was no longer on the point of bankruptcy, the net profit having risen from the disastrous figure of £725 in 1915 to £2,800 in 1916 and £6,000 in 1917. It only remained to be seen if it would be able to rebuild the gun side after all the war contracts ended, and the final accounts were settled by the War Office.

Throughout the war Mabel also had been working hard on various committees devoted to raising money for charities, and for clothing and comforts for the troops. A steady stream of letters from Princess Mary acknowledging her efforts came to her during 1915–18, and she deserved them. But there were some terribly anxious times still ahead before final victory, and Tom and Jim were yet to be in the thick of some of the worst fighting. Jim had gone out to France to join the 12th Lancers in June 1917, and Tom stayed with his brother's regiment as a guest. The 43 Squadron now had Sopwith Camels, but casualties were still very high and a terrible form of cynicism, brought on by the constant loss of friends and comrades, started to set in. In his book *An Airman Marches*, Harold Balfour reproduces a circular he and Tom made up, but never sent out, which would take the place of the letters they were having to write to the relations of casualties, so commonplace had these become.

WORLD WAR 1914–18

The squadron moved to La Gorgue in the north. Passchendaele was over, tanks were newly deployed, and everyone was waiting for the German spring offensive of 1918. Tom flew back to London with Harold Balfour for three days leave and spent the night of 6 January with Jim at Farm House, Old Windsor, when Jim's son, James Oliver Kinloch Rupert Purdey (known for ever after as Jok) was born, father and uncle drinking too much white port in celebration. The following day was Athol's sixtieth birthday, and there was a celebration at the Carlton Hotel, Tom and Harold spending the night at the family flat, 10 Berkeley Street, with Mabel, while Athol was sent to sleep at the Devonshire Club. Next morning Tom and Harold went back to France.

The squadron was very busy and doing very well 'shooting down Huns like partridges', but on 21 March the news came through that the Germans had broken through in the early morning and were throwing back General Gough's Fifth Army in full retreat. Then the whole front was attacked and

Some of Tom's contemporaries during the fighting in France in 1915-18. From left to right: Capt Beauchamp Proctor, VC, DSO, MC, DFC; Major Parker, MBE, AFC; Major Halliday, DSO, MC; Major Maxwell, DFC; Major Fullard, DSO, MC. These officers between them shot down 125 enemy aircraft

the great retreat began. The squadron was moved back to Avesnes le Compte, just behind Arras, and Tom had to get all the ground staff back through the turmoil of a retreating army. He did well, and the squadron fought magnificently, 'both Trollope and Woollett getting 7 Huns in one day each', and the aircraft in the air every hour of daylight. The American airforce, under Major Rice, were operating with them and fought magnificently, 'showing our boys how to do a slow roll', and the news came through of Admiral (later Sir) Roger Keyes's success at Zeebrugge.

On 1 April the name of the Royal Flying Corps was changed to the Royal Air Force and Tom was very pleased; but things were moving fast and at last the Germans were held. And after the battle that saved the British army and France Tom wrote:

> I shall never forget one lovely evening hearing a band of drums and fifes coming down the village street — It was the Guards coming out of the Battle of Arras. What they did in that battle has been written so well by others, but what I saw that night made one feel so proud to belong to the same race as those men. They were terribly few, but to see them march in was an experience I shall never forget. Perfect discipline and turnout, shaved, clean, upright and worn out. I shall never forget that sight.

After this Tom was posted, to his great regret, from 43 Squadron to 10th Wing at Wattignies south of Lille; his wing commander being Louis Strange, DSO, DFC. The tide had turned and the English were advancing. A hundred thousand prisoners a day were being taken and when Tom drove into Lille with the chaplain general the second day after the Hun had pulled out, they were mobbed by the rapturous inhabitants. The chaplain general was smothered in kisses and Tom lost the flashes off his kilt and even the garters off his hose. Two weeks later, at 8am on 11 November, he was woken by his batman with the news that all operations east of the Balloon Line were to cease at 11am. He woke up Louis Strange who laughed; Tom cried, and the war was over.

The immediate aftermath of the fighting ceasing so suddenly was wild enjoyment. The officers found a wonderful billet in Lille and King George accompanied by the Prince of Wales and Prince Albert, the future King George VI, came out to see the troops. There was a gala at the Lille Opera House on 7 December and Tom gave the two princes a lift back to the army commander's house afterwards. But soon everyone just wanted to be demobilised and get home as soon as possible and, as adjutant, Tom found it increasingly hard to organise, discipline and entertain his airmen. In February 1919 the Wing was closed up in Brussels and Tom, now a captain, was sent on further duties to the Rhine; but he sent in his papers in July and returned home to England. The sudden emptiness of normal civilian life after so much activity and responsibility at such a young age made Tom very restless, so before settling down to earning a living he accepted the invitation of a young naval lieutenant, Bruce Fraser, to spend a week as his guest on board the battleship *Resolution* in the Mediterranean. Tom travelled out in January 1920 and had a wonderful time.

Just as he was about to return, a crisis occurred in Constantinople and *Resolution* was ordered to Turkey with Tom on board.

Before leaving Malta, he went to sea in the submarine K12, Commander Ward submerging for Tom's benefit. His great friend Lieutenant Eddie Edmonstone was serving in the destroyer *Vidette* so with the other young officers they picnicked together, bathed and rode donkeys to see the local sights. At Constantinople armed landing parties under Commander Nicholson went ashore to keep order among the local population and Turkish prisoners were taken on board to be shipped back to Malta. Tom found them very pleasant and took photographs of them on the quarterdeck. That episode over, Tom was returned to England, where in June 1920 he joined the firm of his ancestors.

Sadly, Jim's marriage had broken down as so many wartime marriages did, and in 1920 Patsy Purdey obtained a divorce from a badly shell-shocked Jim and took custody of Jok, their only child. She never married again and in later life told her son that his father was the only man she had ever loved but, at the age of nineteen after a wartime marriage straight from the schoolroom, she was completely unable to cope with the awful effects of Jim's shell-shock and war nerves. She brought up Jok in the circle of her family and died on 26 December 1947.

THE TWENTY-YEAR PEACE

In January 1919 Purdey's completed the last military order and the War Office settled its account. After that it was up to the firm to restore its name as gunmakers, by rebuilding its factory on gunmaking lines and finding new markets and customers. Of the craftsmen who had originally joined up, all returned save seven, one of whom had been killed in action, the other six preferring other careers. Throughout the fighting they had been employed behind the lines maintaining and repairing rifles, pistols and other weapons, and their return raised the number of men in the factory to eighty skilled men. Seven apprentices had also left to fight, among them young Harry Lawrence, who joined the Queen's Westminster Rifles in January 1918. He resumed his apprenticeship under the hard tutelage of Alfred Fullalove in the action shop and very nearly left the firm as a result, for Fullalove was a perfectionist and after inspecting his 'boy's' efforts would often fling the work straight across the shop if he was not satisfied with the quality, and would continue to do so until poor Harry was actually in tears! However, Harry persevered and the rigorous discipline of his training laid the foundations of his wonderful skills.

Mr Warner joined the accounts department from the Royal Artillery. He and the first Smith typewriter arrived almost simultaneously in the accounts department; Mr Warner outlasting the machine, still being with us today. The only other major change that had to be made was to remove the women and machinery from 84 Mount Street, so that the premises could be let to Mr Renwick, a dressmaker, in order to bring in extra revenue. The net profit had risen to £12,800 in 1918, but the end of all war work in May 1919, lowered the figure for that year to £7,500, and after a small climb to £8,800 in the following year profits began to decline rapidly until they were a mere £3,700 in 1924.

Athol Purdey was well aware that his guns were a 'luxury' and that war-time increases in taxation were unlikely to be reduced. His English customers, hard hit by the new levels and slow to put in orders, were already complaining at the increase in the price of Purdey guns — in 1920 a gun cost £115, but in 1923 he had to raise the price to £130. This was considered a

Athol c1920

dreadful advance over the prewar price of £95 and, as there were other excellent gunmakers whose guns cost less, Athol did not dare raise the price again for fear of losing too many customers in England.

The trade in repairs and maintenance of existing guns was good, and although orders from Englishmen were scarce, cartridge orders increased as things gradually returned to normal. New guns were ordered from Argentina — four or five a year from Buenos Aires — and Egyptian customers wanted guns for duck shooting. Spain started to re-order on a larger scale, the Duque of Arion, Conde de Bagues and the Duque de Santone ordering during 1921, together with the King and Queen of Serbia, King Carol of Rumania and Prince Sixte de Bourbon Parma. Monsieur Guinard of Paris returned to the scene, but Athol insisted on special terms of payment, demanding one-third on ordering 'and the balance to be paid when the value of the franc reaches 32 per £1 Sterling'. All this was very satisfactory but more orders were needed. The wealth of the United States, comparatively untouched by war, seemed the most likely source of customers in the numbers Athol required, so he planned a selling trip to America for 1922. The Americans, in fact, both individually and through the firm of Van Lengerke and Detmond were showing growing interest, unfortunately most of the orders specifying once again the 'chased breeches and special engraving' so deplored by the Purdeys.

On 12 April, Athol sailed from Southampton aboard the White Star liner *Olympic*. He occupied a state cabin on 'C' deck (£90 one way) and obviously had a wonderful time, as he kept the menus of the dinners he enjoyed with three other passengers of similar tastes who shared a table with him in the main dining-room, and he returned on the same ship after visiting New York, Pittsburg and Chicago. The orders he took were impressive and filled the factory for some months but they were gained at a price of heavy discounts given to both stores and individuals, in some cases as high as 20 per cent, so when the guns were completed after eighteen months work the profit on the trip was very small indeed.

While Athol was trying to drum up support abroad, Tom was doing the same in England. He had no particular title such as 'director', but was just there, learning the business and using contacts he had made during the war. One of the first of these was George Philippi, who in August 1921 paid £378 for a set of three 12-bore guns and in the following year gave a pair of 16-bores to his wife. Both husband and wife were brilliant shots, the former using 18,000 cartridges in 1925. George Philippi had been a hero of Tom's during the war, as he had been one of the original officers under Colonel Smith Barry to help found the School of Special Flying at Gosport with Harold Balfour and Duncan Bell Irving. Another of his wartime mentors was his old commanding officer 'Boom' Trenchard, soon to be

The top two photographs show a Purdey single-barrelled detonating 12-bore shotgun, no 473, built in 1821. The stock is made of bird's eye maple wood. Weight 6lb 5oz; barrel length 30in

The third photograph shows the detail of a Purdey 12-bore pinfire double-barrelled game gun, no 6829, built in 1864. Weight 6lb 14oz; barrel length 30in

The bottom photograph shows a Purdey 12-bore over-and-under game gun built for the USA in 1983. Weight 7lb 5oz; barrel length 28in

Richard Beaumont in the Long Room in 1984

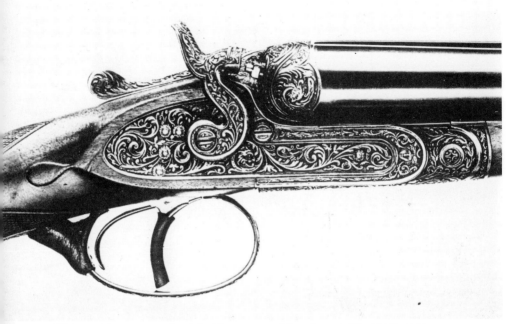

The special gun put up by Athol for a competition shoot in Madrid in 1921

first Marshal of the RAF and a viscount, who brought his Lancaster gun for servicing and remained as a customer.

Tom was also very keen to arouse popular interest again in the shooting grounds. During the war years the attendance had suffered badly, with sixty customers attending in 1915, forty-three during 1916 and only thirty in 1917. He therefore set himself the task of learning to shoot well and at the same time tried to make up for the lost war years by going to the Purdey factory and learning as much as he could about the making of guns, although he was very indignant when one of the old actioners told him that his efforts to use a file reminded him of 'a cow trying to handle a musket'.

Jim, on the other hand had, after his divorce, been appointed as an ADC to the Governor of Dar es Salaam, Sir Norman Byatt, and while there he met Bee Oliver whom he married on 22 September 1923. Miss Oliver was related to the family who controlled Debenham and Freebody and the relationship between her family and Jim was to have a profound effect on the firm of Purdey in years to come.

In spite of the difficult financial climate caused by taxation, the English market started to improve. In December 1922 Lord Louis Mountbatten bought a pair of 12-bore guns and Lady Louis a 16-bore. Lord Lascelles or-

Lord Lascelles and Princess Mary shooting grouse with Lord Lonsdale

dered a 12-bore and Princess Mary gave him two more to make up a set of three. Prince Obolensky, then living at Hanover Lodge in Regent's Park, had a pair; and the King of Rumania ordered the same. As shoots got going again, cartridge orders increased and men were taken on to load them, extra men being also hired for the shooting grounds, and orders started to increase from France, Spain and Italy. The King of Spain ordered a set of four 14-bores for shooting partridges; and in 1924 a pair of 12-bores followed by a pair of 20-bores for his eldest son, the Prince of the Astorias. General A. Canton de Wiant, VC, 'the most wounded man still alive', as he called himself, came in 1923 for a gun. Japan ordered fifteen, Australia four and Manton of Calcutta eight rifles of .369, .246 and .465 calibre.

Tom, at the same time, lived up to the reputation of his ancestors by chivvying the customers who did not pay their bills, and added his own form of comment to the ledgers. 'Disgusting' he writes beside a doctor who had been out to the Sudan for four years and did not settle his account. He went further in 1926: 'Bloody bad. Tried to BORROW! Too Awful. Never lend anything.' Against a major: 'Don't let this man have *anything*.' To a third: 'Stinking Bad and a Shit 1936'; followed by 'Still is, Jan 13/37', and 'More so; Feb. '38', against the name of a city merchant who lived in Half Moon Street. As Leonard Lowe, in the accounts office, pointed out, this was all very well until a customer demanded to see his account in the ledger, when there were frantic rubbings out down in the office before the book could be brought up to the Long Room!

Among the guns that were produced in 1923 were a very special pair built for Queen Mary. The Queen's Dolls' House, now on exhibition at Windsor Castle, was to contain everything that a great house should have and naturally, as King George V was such a keen sportsman, a pair of guns had to be included. A perfect pair of hammerless side-by-side guns were, therefore, made by Harry Lawrence (actioner), 'Gus' Schakel (stocker) and 'Mickey' Miles (finisher). Harry made the actions and fore-ends, and assembled the guns. They were too small to actually shoot, being only four inches in length when assembled, small pieces of india-rubber taking the place of the top-lever springs which close the bolt. The guns in their case were so charming that instead of being put in the Dolls' House gunroom, they were placed by the King on the writing desk in the Dolls' House study. The actual entry in the ledger reads:

August 1923.   Her Majesty Queen Mary.
A Pair of Miniature Guns Nos. 22491/2, with Case, Fittings, Shooting Seat (King's Pattern) Cartridge Magazine and Cartridges.

They were presented to the Queen by Athol as a gift from his firm. Sir Edwin Lutyens, the designer of the Dolls' House, wrote to Athol on 21 April 1923 saying that the photographs of the guns were delightful and would he bring Mabel to see the house which at that time was in pieces, but would be together in three weeks time.

Among the other events of that year was the death of an old and respected customer, Mr Vincent Corbett JP, who made special reference in his will to the disposal of his Purdey gun, and Athol kept the newspaper account of his bequests firmly stuck alongside his name and last entry in the appropriate ledger:

Mr. Vincent Charles Stuart Wortley Corbett, J.P., of Chilton Moor House, Fence Houses, Durham. Mining engineer, a Director of the Londonderry Collieries Ltd., and for 50 years Chief Mining Agent to the Londonderry Collieries, who died on November 28, aged 89, son of the late General Sir Stuart Corbett, left estate to the gross value of £47,036. with net personalty £42,678. Testator directed that:-
He should be buried with his first wife in Kirknewton Churchyard, that

his body should be conveyed from his residence to Chilton Moor Church, and then to Fence Houses Station, on a trolley drawn by his old roan horse, and from Kirknewton Station in any manner found convenient, except that a hearse may not be used.
He left:-
To Mark Harrison, gamekeeper at Plain Pit, his gun [sic] (except his Purdey gun, which he left to his son-in-law, the Rev. Morris Maurice Piddocke) and rifles, his sporting dogs and binoculars, and £100; his Poultry, Swans, and White Muscovy Ducks to his wife; his fishing rods and tackle equally between his sons-in-law, the Rev. Morris Maurice Piddocke, Bernard Gilpin Forman, and John Alexander Nellan; £200 to his secretary, Joseph William Hall, and £100 to each other executor; £70 each to his groom, Robert Jackson, and his gardener, Joseph Hume, and £20 to his gardener's assistant Lizzie Stokoe, if respectively still in his service; £25 to each other domestic servant; £20 to Thomas Mark Harrison, gamekeeper at Cocken; £25 to William Newlands, shepherd at Kirknewton; to "Mistress" Anderson (widow of Robert Anderson) £25 and his effects at 38, Church Street, Seaham Harbour; £250 to Malcolm Dillon as a momento [sic] of his kindness; £10 for a momento [sic] each to Anthony Scott, A. B. Hare and Asquith Swallow, if still associated with the colliery undertakings with which he was connected.

Tom, meantime, still felt the urge to travel, so in the company of Sir Archibald Edmonstone he went all the way to India in February 1923 because his old friend Eddie Edmonstone, who had now joined the cruiser *Colombo* on the India station, had been granted fourteen days leave. They sailed in the ss *Caledonia* by way of the Suez Canal and Aden, and joined Eddie at Bombay. They saw Lutyens' Viceroy's House at New Delhi only half built and visited the Taj Mahal — which Tom described as looking like the Pavilion at Brighton — Agra, Fatehpur Sikri and Bombay where they stayed with the governor, Sir George Lloyd, before returning on the *Caledonia* on 3 March. It was an expensive trip, but the firm was completely a Purdey concern, and the members of the family considered the revenue to be entirely their own prerogative. At that time the family joke was the division of the yearly profits between the two partners, Athol and Cecil, whom Mabel most unfairly called 'Uncle Swizzle'. According to Purdey family lore, Athol and Cecil would lock themselves into a room at the end of the financial year on 1 January. Through the keyhole could be heard Athol's voice intoning, 'Two for me one for you; two for me, one for you', as he distributed the pound notes from a huge stack of money at his elbow. In fact Cecil was taking only his salary from the business, and

Queen Mary's Dolls' House, showing the miniature guns in the study always repaid any loan as soon as possible. He had a genius for model-engine building and with the help of Harry Lawrence built five working steam engines which were used at the miniature railway at Purley and were exact replicas, in miniature, of existing locomotives.

In the meantime, in spite of new orders the profits of the firm continued to fall keeping the low figure of £3,789 in 1925. Jim and his new wife had returned from Dar es Salaam, and set up house in London. He looked everywhere for a job but was unable to find one. Mr F. S. Oliver, his wife's father, then suggested that, as Jim loved the land, he would arrange for him to learn farming near Jedburgh. This was a disaster as Jim's idea of farming was hunting, shooting and a good farm-manager, none of which he could afford. Mr Oliver next approached Athol and Tom with the proposition that if he were to put the money he controlled in Bee's trust into Purdey's, could Jim become a director. The Purdeys agreed and Jim joined the family firm. The Olivers believed that Mabel's extravagance was one of the main reasons for the perilous state of the Purdey fortunes and felt very sorry for Athol. Mr Oliver could not bear Mabel and his prejudice was confirmed when he heard from his daughter that, even in those hard times, Mabel could still insist 'Well I have always said only the best is good enough for me.'

The factory was full of orders and everything should have been sound but the costings were wrong and outside money had to be found to put the firm back on a firm financial footing. Jim's Oliver brothers-in-law, with their resources and organising ability, were ready at hand to help. The business therefore had to be turned into a limited company, with preference shares to protect the money supplied by the investors and the position of the family, and ordinary shares also available. The old order was to go; efficiency was to be king, and after these modern changes the firm would

Athol shooting pigeons in Monte Carlo in 1923

prosper as never before. 'Efficiency', however, had never before come face to face with either Purdeys or English craftsmen, who worked in their own way and at their own speed. Athol, who knew his business inside out, found himself pressed to go along with the new ideas put forward by his sons who had not yet learned the basis of their trade or understood the financial limitations of the company. It was not so much efficiency that was needed as economy and prudence, but these were heady days after the restraints and disruptions of the war years, and the younger Purdeys felt that their father's and uncle's cautious views were not in keeping with the times.

The first meeting of James Purdey and Sons Limited took place in the Long Room on 5 October 1925. Athol was chairman and Cecil, Jim and Tom Purdey accepted, and paid for, preference shares and became directors. Jim Purdey's Oliver relations bought cumulative preference shares in the business. Athol was appointed managing director at a salary of £1,250 per annum, Tom became manager at a salary of £350, and Jim a director at £250, on the understanding that all three should devote their whole time to the service of the company. George Tillson the head clerk who had been with the firm all his working life and was then sixty-seven years old was made secretary to the company. After this formal meeting new ideas started to be introduced, Tom Purdey being put in charge of alterations in the methods of accounting and bookkeeping to make them more efficient. Jim Purdey, told to find an advertising expert with a view to developing a definite advertisement policy, employed the services of Messrs Sampson Low. At the next board meeting Tom announced that his exercise on efficiency would entail the employment of another clerk in order to make his system work. The new clerk was Graham Tollett. On joining the firm, his first job every morning was to take up a new bottle of whisky to the Long Room for Tom and Jim, and after that to get on with his work. One of the most loyal and conscientious of men, he was eventually responsible for all the correspondence of the firm until his retirement in 1976. Tom was made an assistant manager and assistant company secretary in 1926. Jim was made an assistant manager and assistant company secretary as well, with a salary for each post. But more orders were still needed to keep going, and Tom prepared for a tour of America during April 1926. Tom's visit to the United States was a remarkable success, for he took more than forty orders for guns and rifles. Unfortunately Jim's efforts were not so well received, his scheme of advertisement being so expensive that it was turned down as being beyond the company's means.

Social conditions in England were not at all conducive to good trade for a firm like Purdey's. The state of industry and the mining areas was such that, after great unrest, the country suddenly found itself facing the fact of a general strike from 3-12 May 1926. Purdey's factory was unaffected by the prevailing mood, and when transport in the metropolis came to a halt the men whose homes were too far away simply spent the nights on mattresses and blankets on the floor and carried on as usual. Tom and Jim volunteered to help keep public services going, and for the first two days of the strike the two brothers worked as porters at Paddington station under the direction of an army sergeant. All went well until Tom was detailed to help unload a milk train. He was positioned at the bottom of the ramp and the first six milk churns were empties. The rest were full but the man on the train omitted to warn Tom of this fact, with the result that he was knocked over and the whole contents spilled on the platform. The sergeant was furious and Tom was removed to the meat-assembly point in Hyde Park, where he contracted awful boils on his neck from carrying frozen meat on his back. He was further incensed when he overheard the men in the factory suggesting that his boils came from a very different activity, and one of his famous rages ensued.

Jim, after completing his stint as a porter, joined the Special Constabulary, mounted section. His experiences were most unpleasant and he was shocked by a confrontation which took place between militant strikers and mounted specials near Marble Arch, the idea of taking action against his fellow Englishmen being particularly repellent to him. Graham Tollett, the new clerk, owned a motor bicycle, so when he also volunteered as a special constable he became a despatch rider at the mobile section at Marylebone Police Station. It was part of his duties to take official documents to police and army headquarters in London, and he was present at the incident near Marble Arch in which Jim was involved. It was so unnerving that he was one of the first to take cover from the brickbats and violence. The mood of the workers however was ambivalent, and Graham was surprised by one incident when he was part of a police escort for a convoy to the docks. A crowd of strikers barred their way and things looked very ugly, but while both sides were waiting to see what would happen one of the strikers said, 'What about a fag?' A police special constable, a Harley Street doctor, handed over his gold cigarette case. Graham thought the case had gone for ever, but when all the contents had been taken it was handed back to the owner and the convoy went through.

The aftermath of the strike was a period of uncertainty and bitterness throughout the land; but the gun trade was not affected, and the boom in the United States was ensuring that a constant flow of orders was coming into the factory. After the immediate effects of the strike had died down, English customers continued to send in their guns for servicing and the number of cartridge orders started to increase. The latter were still being loaded by hand to the individual's selection, and the numbers ordered were to increase steadily until yearly amounts of 15,000 to 20,000 per customer were commonplace. The cartridge shop was run by Bill Leeper, a small and charming man who worked in Purdey's until 1952. His father,

William Leeper, had been a shipwright at Southend, and when James the Younger put in there for repairs in his yacht *Lynx* in 1872 his carpentry made such a good impression on the Governor that the latter offered him a job in Purdey's doing repairs on the buildings and making boxes for cartridges. He accepted the offer and was installed in the basement but James did not know that his hobby was building violins, and customers in the front shop were amazed to hear melodious strains coming from beneath their feet as old Leeper tried out his products. Now his son Bill was in charge of the loading of all cartridges for the firm, with Ralph Lawrence, the second son of Ernest Lawrence, as one of his assistants. Ralph had great ability and introduced a method of mass production to the loading system, thus increasing the numbers of cartridges loaded and supplied each year.

Guinard of Paris ordered a pigeon gun for El Glaovi the great Marrakesh leader, and a pair of guns for His Majesty Maulay Hafid, King of Morocco, both guns being built with hammers for pigeon shooting. Sir Ian Walker-Okeover, one of the great shots of the time, ordered a set of three 12-bores in 1930, and his records of grouse bags at Milden in Forfarshire became legendary. The King of Serbia ordered one more gun and the King of Spain a .246 rifle, but the firm of Eduardo Schilling of Barcelona, who had helped so much in the war, went bankrupt. At the same time the bad debt position of Purdey's was worsening steadily, due to the laxness that had crept in over chasing customers who would not pay, the younger Purdeys fearing to antagonise customers who were obviously making the best out of this situation.

In the short term, new orders and reorganisation of the Audley House staff had some effect as the net profit for 1927 rose to £10,000, the highest figure reached since the days of the 'old Governor', and as a result salaries could be raised. In addition, the freehold of a shooting ground was bought at Eastcote, South Harrow, the first freehold property owned by the firm. The new ground had one unfortunate disadvantage in that it was situated close to the local sewage works, and when the wind was in the wrong quarter the smell was unbearable. Hennington, the man in charge, was in daily touch with the front shop and, in answer to Athol's telephone call as to what it was like at Eastcote today, was apt to answer, 'Don't send anyone up today sir. The smell is awful. The manager of the sewage works must be stock taking.'

The position of head coach was eventually taken over by Bill Morgan, who came to Purdey's from Lancaster's after that firm ceased to trade in 1933. His job was to test all new guns and rifles at the grounds before they were finally passed by Athol in the Long Room, and also to coach the customers in grouse shooting at which he was outstandingly clever. His stories were, however, difficult to believe, in particular his claim to have been the only soldier in the Dardanelles campaign who, while on sentry duty, bagged a Turk with one shot from his rifle, reloaded and then shot a woodcock which had been disturbed by the sound of the report. His instructions to his clients were invariably couched in terms of great respect, unlike the more blunt approach from Harry Smith when the two were joint instructors at the West London Shooting Grounds at a later date. 'Follow through nicely sir, if you wouldn't mind', was Morgan's tactic. 'Pretend he's got smoke coming out of his backside. Now Your Grace, follow up his smoke trail', Smithie would say. Both systems apparently were equally effective in the long run.

But to return to the problem of encouraging more customers, as Jim had been in India in the early part of the war it was thought good policy to use his experience and contacts by sending him out on a selling trip to meet and stay with the maharajahs. He therefore travelled round that country for two months during 1928, taking many orders for rifles, while Tom went off to America to take advantage of the continuing US boom. An added incentive to their working hard was that on these occasions they were both granted a 5 per cent bonus on all guns they managed to sell. Tom took many orders, granting discounts of 15 per cent and extra for cash to Abercrombie and Fitch of New York, with the result that more men were taken on in the factory in October 1928, and more apprentices were recruited. As well as the special bonuses for their efforts abroad, a dividend of 22½ per cent was paid on Tom and Jim's ordinary shares and their salaries were raised by £250 a year. Dinners were given in the Long Room on every conceivable occasion: Derby Day, Grand National Day, Armistice Day, before Ascot, after Ascot, Christmas Lunch etc. It was all great fun and the firm's outings and 'Old Comrades' dinners were celebrated in style. Everyone felt that wonderful times were here for ever.

Frederick Beesley, the inventor of the Purdey action for the Governor, died on 14 January 1928, in his eighty-second year. Having started work as an apprentice to Moore and Grey in 1861 at the age of fifteen, he had risen to be one of the most respected gunmakers in London, and his last invention when he was seventy-seven years old was described in the *Field* of 24 May 1923 as a 'marvel of ingenuity'. It was a unique design to give a straight blow to the strikers of over-and-under guns.

In March 1928, Athol attended a dinner at the Holborn Restaurant in honour of his seventieth birthday. Everyone from the factory, front shop and shooting grounds, with the exception of apprentices, was invited — a total of eighty-seven men, and Athol presented each man with a silver pencil-holder engraved with the date and his initials. At the end of dinner the whole company rose unsteadily to their feet, the door was opened and Mabel entered to hear the speeches. She walked sedately up the side of the room and at that moment Gorman, one of the cartridge loaders, rolled slowly out from underneath the table straight into her path. Without

Athol with Jim and Tom in the Long Room in 1928. On the mantelpiece can be seen photographs of James I and James the Younger, and cartoons of Lord Henry Bentinck

moving a muscle of her face she gathered her hocks beneath her and cleared him at one bound and then moved on; nobody even dared to smile.

In 1929, King George V became very ill, but during his convalescence he felt he must exercise his arms otherwise he might lose his swing and sense of timing; however, he was extremely safety minded and would not allow a gun in any room in the house apart from the gunroom. Athol and Harry Lawrence overcame his problem by building an exact replica of the King's guns in wood — the same stock measurements, two dummy hammers and hollow wooden barrels. The weight and balance were the same, but inside the barrels of his imitation gun were placed bulbs and batteries with small fitted lenses so that when the triggers were pulled a beam of light was projected onto the object aimed at. This idea worked perfectly and His Majesty was able to practise in his rooms until he was well enough to resume normal shooting.

As a result, the try-gun in the Long Room was fitted with the same electrical appliances, and customers have been fitted for their guns with this system ever since. A white card bearing the silhouette of a flying pheasant has been fitted to the wall of the Long Room and, after his measurements have been placed on the try-gun, the customer can see by the projected light exactly where he is aiming. The added advantage is that older customers who have been shooting badly and cannot understand why their shooting skills have deteriorated so much, can be shown where they are aiming. As one gets older the left eye usually tends to take over as the 'master eye' and, should one shoot with both eyes open, will pull the barrels to the left of the target. The solution is either to shut the left eye or to apply cast-off to the gun to compensate for this left 'pull'. At one time no boy was allowed to close his left eye as it was said 'it was not the way to shoot'. As a result many boys who had a strong left eye had miserable days of shooting as they kept missing to the left. This fashion was handed down from the days of Lord Ripon, who had perfectly balanced eyes so did not need to shut his left eye or have cast-off on his guns. As he shot so well, people thought that to do the same all one had to do was to copy his every act, but as each individual has a different breadth of shoulder and his own standard of eyesight, the fitting of the gun will be different in each case. Another fad was to urge young shooters to shoot fast. This again came from those same days when brilliant shots, who did little else but shoot from 12 August to 1 February each season, naturally reached such a skill that they could increase their speeds of firing with accuracy. Everyone again has his own speed of reaction or swing, so the speed at which a boy can shoot and hit his target should be entirely his own affair and pleasure.

The Duke of York, the future King George VI, had a pair of light 12-bore hammerless guns finished for him in 1930, and the Duke of Kent had a similar pair in 1931. The King, still weak from his illness, had a new pair of light 12-bore guns built, with hammers, and was so pleased with their performance that he added a third later in 1931. All the Royal brothers — the Prince of Wales and the Dukes of York, Gloucester and Kent — were keen shots, the Duke of York becoming one of the best shots of his generation. They corresponded frequently with Tom and Jim, always writing in their own hand on any subject to do with their guns. Tom lent his own guns to the Duke of York and the Duke of Kent while their personal guns were being finished in 1929 and 1931 respectively (see letters overleaf), and in 1922 the Duke of Gloucester had written from Canterbury where he was stationed with the 10th Hussars:

> Dear Tom,
> Will you please do me a great favour and let me shoot at your grounds next Saturday morning. There are three reasons:
> 1 I cannot manage an earlier date.
> 2 I want some practice.
> 3 I want my servant to learn how to load.

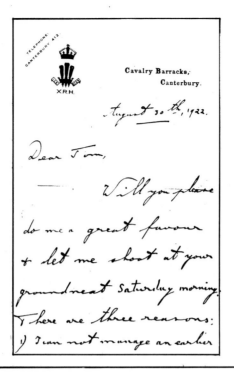

> Cavalry Barracks,
> Canterbury.
>
> August 30th, 1922.
>
> Dear Tom,
>
> Will you please do me a great favour & let me shoot at your ground next Saturday morning. There are three reasons:
>
> 1) I cannot manage an earlier
> 2) I want some practice
> 3) I want my servant to learn how to load.
>
> I shall want to take my guns away with me on Saturday or if the shoot is impossible will you please have them sent to Buckingham Palace on Friday. I am sorry that the notice is so short.
>
> yours sincerely
> Henry

His next letter was more practical: 'Please send receipt and any future bills direct to me and not the Comptroller (he does not exist!!!).'

A.

145 PICCADILLY
W.1.

October 22nd 1929
—

Dear Purdey,

Thank you very much for lending me your own guns for this week. Are you quite sure you do not need them? I will take great care of them. It will be very interesting to see what difference there will be in the results. I will give them a good trial. If you will put them on the train I told you in charge of the guard I will have them met. Many thanks

Yours sincerely

Albert

BALMORAL CASTLE.

August 30. 1931.

Dear Tom.

Thank you so much for sending me a pair of 12 bores & I am sorry there was a muddle over them & I hope it didn't inconvenience you. I thought when I asked for my guns to go to Sandringham that the 12 bores you lent me last season would be sent, not realising that you might be selling them. Anyway, this pair are very good, not that I have hit many grouse so far — I never can hit anything with 16 bores & so I will come & see you about them when I get back — I hope you haven't had too bad a time in the way of gun-selling etc.

Again very many thanks
Yrs. G. Sincerely,
Troyte

By 1929 Jim was divorced from his second wife, and he and Tom went together to America. While they were there Jim met Mary Stuart La Boyteaux, younger daughter of William Stuart La Boyteaux of Greenfields, New Jersey. They were married on 28 August 1930 in Riverfields, New Jersey, and this marriage was to be an outstanding success. The couple returned to England to a house in London, 40 Ovington Square, changing later to 5 Wellington Square, Chelsea, where their only child,

Jim and Mary La Boyteaux on their honeymoon, 1930

William, was born in 1937. They then moved to Farley, close to the golf course at Wentworth.

But in 1930 events outside England were about to change the prosperous and happy times the Purdey brothers had been enjoying. The last trip to America had not been a success; there was caution regarding the future and fewer orders were taken. The profit position reflected this change with a figure exactly half that of the previous year and in 1930, when with so few orders coming in the factory had to revert to the building of guns for stock with rough-hewn butts which could be altered to a customer's measurements for immediate sale, the figure fell to £3,426. The following year, 1931, there was an overall loss of £500. The American market had virtually ceased to exist; in 1930 there were orders for 64 new guns but the figure was to drop to 23 in 1932 and to a mere 6 in 1934. The Depression had arrived and drastic action had to be taken to save the firm from bankruptcy.

The preference shareholders, the Olivers, who were in a difficult position after their sister's divorce, still wanted to safeguard their investment, and they were voted onto the board at a meeting in April 1932. Quite rightly, they stipulated that the cash position of the company be closely examined at each board meeting and reconciled with the firm's statements of expenditure. Letters were to be written to debtors who had not replied to earlier appeals, and the solicitors were to effect collection where necessary. At Audley House the strictest economy was to be exercised and *all* entertaining was to be abolished. In order to show how serious they were in their demands, at the next board meeting in May 1932 Cecil Purdey staged what can only be described as a 'palace revolution'. Taking the chair in the absence of his brother Athol, and representing the preference shareholders, he read a memorandum dealing with the past administration of the company's affairs, and demanded the most rigid economy and reduction of expenses both in the factory and at Audley House, especially with regard to the directors' remuneration and expenses. He backed this up by laying particular stress on the need for the closest co-operation between the directors — a shrewd dig at the way in which he had been taken for granted — also stating that the indebtedness of Athol, Tom and Jim to the firm should be dealt with at once. By this time Tom owed the firm over £1,300, with Athol and Jim each owing about half that sum. Not only had their salaries and bonuses been high, but almost all their personal expenses had also been put down to the firm each year. This was now to be stopped, and while no dividends would be paid, the firm would have a lien on all the shares held in trust for both Athol and Tom. Cuts were to be made in the salaries of the clerical staff and, most important, only ten more stock guns were to be made as people were not buying them at that time, the inevitable result of this being the reduction in the number of men in the factory until trade improved. Various other cuts in directors' 'perks' were mooted, but in order to soften the blow slightly for Athol and Tom they could occupy rooms on the third and fourth floors of Audley House free of rent.

All this was a terrible blow to Athol. He had bought a house, 53 Earl's Avenue, in Folkestone, and here he now decided to live while Mabel moved into the third-floor flat over Purdey's, next to Tom who was in the flat on the fourth floor adjoining the cartridge-loading shop. From this time on, Athol virtually ceased to have any say in the running of the business. Much against his better judgement he had handed over control to Tom, and only came up to the shop for board meetings. The Long Room was being reorganised, and the white pheasant given to James the Younger by King Edward VII was moved out into the hall. Eugene Warner found Athol standing in front of the glass case gazing at the bird and heard him mutter: 'So they have thrown you out as well!' In April 1934 the board voted him an annual allowance for life of £174 in consideration of past services.

Having taken the decision to cut down on the work force in the factory, Ernest Lawrence was summoned to the Long Room in May 1932 to discuss the best way in which this could be carried out. It was decided that twenty-six trained men should be dismissed. As no pensions were then available it was thought better to hold on to the older men who would not be able to get other jobs, and to get rid of those in the middle age-group, as with their skills it was more likely they would find other employment. Lawrence was told to go back to the factory and work out who should stay in each department and who should go. He consulted his son Harry, who pointed out that should the recession last for any length of time, when it was over the older men would be wanting to retire and there would be no one available to fill the gap of training the apprentices for the future. As a result the board altered its decision, and decided to dismiss twenty-six of the older and more experienced men over the age of fifty-five, so that the younger men could maintain the continuity of the firm. And however harsh this may seem, this action saved the firm for the future.

In the factory, feeling was very bitter; for these older men had no prospects of getting work outside as other gunmakers, in even worse straits, were cutting down on staff as well. One week's wages were given to each man, and Lawrence and his son Harry had the unpleasant task of carrying out the sackings. The dismissals were stretched over all departments. In the barrel shop one man had to go; whereas six actioners including Alf Fullalove, Harry's 'gaffer', were dismissed. Four went from the ejector shop, as well as three stockers, two finishers, two from the machine room and no less than six repair men. On top of that, two staff from the shooting grounds went including Mr Sam Smallwood, a senior instructor of great experience and much respected in the firm. Only twelve returned two years later when trade and orders eventually picked up once again, and another three by 1936.

Ernest Lawrence, factory manager, and Tom, on a firm's outing in the 1930s

An ironical twist to this situation arose during the high inflationary years of the 1970s. At that time the price of a new gun was increasing rapidly and the delivery time was three and a half to four years. Prices for second-hand Purdey guns with quick delivery were, therefore, proportionately high from dealers and auction houses. Our customers told us that certain such sellers had informed them that the 'vintage years' of Purdey craftsmanship were 1926, 1936 and other specified years between. This made no sense to the men in the factory, ten of whom were still with us having joined in 1927 and 1929 and had put their initials on each piece of work they produced. Their 'gaffers' did not, they pointed out, produce their best work in 1926, allow their quality to drop for a year or two, then do their best work again in 1936. Neither did they themselves, as they strived to produce better work in each consecutive year. Further to that, Ernest Lawrence and his son Harry were very much in control of the manufacture of each gun for each individual customer, and it is most unlikely that they would have passed any gun which was not up to their searching standards. This talk of the vintage years being in the 1930s therefore makes no sense, especially in view of the dismissal of twenty-six of some of the firm's most experienced craftsmen at that time.

But to return to 1932, and to Tom who was looking ahead to the better days that must eventually come. The Silver Jubilee of King George V was to take place in May 1935, and he decided to give a very special present from the firm to the man for whom he had such great respect. Knowing that the King had been so pleased with the pair of miniature guns that had been given to Queen Mary for the Queen's Dolls' House, he decided to build for His Majesty a pair of miniature hammer guns in all respects exactly the same as the guns used by the King but one-sixth of their size. Unlike the guns in the Dolls' House, the new miniature pair were actually capable of firing. Harry Lawrence was given the task of making them and so, with the assistance of his youngest brother Ernest, then aged twenty-two, he started work on what must be a masterpiece of gunmaking. The guns, Nos 25000-1, were 7in long when finished, they each weighed 13 drams and were indeed perfect replicas in every detail of the King's real ones. The hammers when pulled back to the cocking position were retained by tiny replica locks, and leaf-springs activated the firing action. The King's insignia was let into the stock of each and the cartridges, containing 1.62 grains of EC powder and 2.02 grains of dust shot, were loaded by Imperial Chemical Industries at their Eley plant in Birmingham.

In case there was an accident in the building of one gun, Harry made three, and these three guns took three years to complete. When finished, a gold and silver double, flat case was made by the jewellery firm Garrard of Regent Street, and Harry then made a complete set of miniature fittings, turnscrews, oil-bottles, snap caps etc to go in this case. The two guns were then taken to Buckingham Palace by Tom in the Jubilee Year and presented to the King. The third gun, No 24707, was kept by Tom Purdey. His initials were engraved on a gold oval in the stock and, with a leather case made by Harry Lawrence's mother, this gun was presented to Tom by the Lawrence family on 22 March 1938, on the occasion of his forty-first birthday. Tom was invited to the Jubilee Service in St Paul's Cathedral on 6 May and sat in seat 11, block 'C' south nave, wearing the full-dress uniform of the Argyll and Sutherland Highlanders, claymore and all. And his uncle Cecil, who was Master of the Gunmakers' Company in that year, at the invitation of the Lord Mayor of London was present at a reception, ball and concert held in the Guildhall on Wednesday 22 May. The famous Purdey decoration was taken out of the basement and erected on the front of Audley House, and the whole building was draped in flags and red, white and blue bunting.

As their own personal celebration, the entire firm went by train to the Albion Hotel at Worthing. After a six-course dinner — salmon, cutlets, chicken etc — came the toast 'the King', followed by 'our Army, Navy and Air Force', then 'the Firm', 'Captain James Purdey and Mr Tom Purdey', 'the Trigger' (the traditional gunmakers' toast) and 'the Chairman', who was a man elected from the factory. In between these toasts eleven different songs, recitations, poems and turns were performed by those individuals

Harry Lawrence with one of the miniature working guns he built as a gift for King George V on the occasion of his Silver Jubilee in 1935

brave enough or still sober enough to stand up, and the evening ended in a general sing-song and drinks party.

Harry Lawrence was also employed in building and perfecting the over-and-under gun. Although Purdey's had built over-and-under rifles as long ago as the 1870s, Athol and his father had concentrated on the side-by-side gun and did not want to become involved in guns of the other design. However, during his visit to America in 1920, Athol had been approached by several customers wanting such guns and citing the fact that those made by both Woodward and Boss were having great success on the Continent and in the United States. As a result Athol decided to build over-and-under shotguns, having seen a gun by Green of Cheltenham which he considered had strength and a strong action. Up to that time he had been prejudiced against guns of this design as he had reservations as to their safety, but this gun had six bolts and this convinced him that the barrels would be held tight on the face of the action. The first Purdey prototype consisted of old useless barrels brazed together and thickened with extra metal 'tinned on' to make up the size at the breech end, and a trial action machined in the basement of South Audley Street. This first mock-up was never fired or proofed, but gave Harry Lawrence the experience he needed when the first barrel tubes arrived from Vicker's. He then built a complete gun which, owing to the nature of the six bolts, weighed 8lb 2oz, which of course was far too heavy for ordinary use. It was suggested to Athol that the top bolts were unnecessary and weight could be saved by discarding them, but to him safety was all-important, so two or three of these very heavy guns were built each year during the next three years. After Athol's retirement, Harry persuaded Tom to do away with the top bolts leaving only four grips instead of six. The action was modified accordingly and the weight reduced to approximately 7lb 10oz for a 12-bore gun with 30in barrels, ten of these guns being built during the thirties and, by the outbreak of World War II, eighteen over-and-under guns on the Green principle had been made and delivered, all having been designed and actioned by Harry Lawrence with the assistance of Len Howard, a side-by-side actioner.

The story of the development of these guns did not end there. During the war years none were constructed and it was not until 1948, after Tom Purdey had returned from his first postwar trip to America where he again received several requests for guns of this type, that manufacture was resumed. The guns were to be of the old design, namely having four bolts but, by a strange coincidence, Charles Woodward of the famous firm of James Woodward and Sons who manufactured over-and-unders, approached Purdey's in the summer of that year and asked Tom if he would consider buying the Woodward business. The Purdey board, after studying the records and accounts, decided that no offer could be made as there were only two or three good craftsmen and the profits were virtually non-existent. On hearing this decision, Mr Woodward announced that he could not bear the thought of his business being handled by anyone except Purdey's and would like Tom to take the whole concern off his hands. As a result, a contract was signed between the two businesses in September 1948 for 'the

Five grand old customers in 1933. From left to right: Capt W. H. Burn (80 years); The Earl of Southesk (80 years); W. Shaw Adamson (83 years); Col W. Steuart Fotheringham (71 years); Major Lindsay Carnegie (82 years). On 16 December 1933, when their combined ages equalled 396 years, they shot 114 Pheasants, 9 Hares, 1 Teal, 1 Snipe, 3 Partridges and 36 Rabbits

purchase of the goodwill and use of the name of Messrs. James Woodward and Sons of 29 Bury Street, St. James's, W1 for the sum of £300, and for the purchase of materials for the manufacture of the Woodward over-and-under gun to the value of £144 14s 6d'. A commission of 10 per cent for five years was to be paid to Charles Woodward on orders for cartridges and repairs as per a list of approved Woodward customers.

As soon as this transaction was completed it was decided to abandon the old designs and build the over-and-under gun under the name of James Purdey and Sons, but to the design and ejector mechanism of the Woodward gun. As Purdey's built side-by-side guns on their own action, the Woodward side-by-side gun ceased to be constructed from that date onwards. Ernest Lawrence, who had returned to the firm after war service in the Royal Navy, now took over the building of all over-and-under weapons and with George Wood, barrel maker, made slight changes and modifications in the look and design of the guns as time went by (see Appendix 2).

In 1948, Mr Joseph Nickerson (now Sir Joseph) ordered the first Magnum 12-bore over-and-under. Built with 28in barrels this gun, No

26113, was delivered in 1950 and weighed 7lb 12oz. Sir Joseph is one of the few English customers of ours to shoot game and wildfowl with guns of this design and has trios of them in 12-bore, 20-bore and 28-bore sizes. He shoots brilliantly and his great bags of wild greylegged partridge in Lincolnshire, and grouse scores at Wemmergill in Yorkshire, have all been achieved with over-and-under guns of our make. From Harry Lawrence's first Purdey over-and-under gun in 1925, a 12-bore with 31in barrels and weighing 8lb 2oz, 266 such guns of all bore sizes had been built by the end of 1983, by which time the weight of a 12-bore with 28in barrels and chambered for 2¾in (70mm) cartridges was down to 7lb 4oz.

However, while the 1930s development and improvement of over-and-under guns was taking place, the main orders for guns were slowly improving — 157 in 1933 and 180 in 1934. The stringent cutbacks of 1932 were having a satisfactory effect on the finances of the company, and by the end of 1933 the firm was making a small profit. Tom even took out a provisional patent for a gun fitted with electric lights in the barrels, known as the 'flash' gun, and 'cartridges' which had bulbs and batteries fitted into a tube the size of a 12-bore which could be used in an ordinary gun. Another of Tom's inventions was the 'egg bomb'. There had been a spate of bank robberies and Tom thought up the idea of splashing the offenders, or their getaway cars, with paint which would make them easier to identify and apprehend. Several hundred small clay balls about the size of a cricket ball and filled with sticky blue and yellow paint were made up in the factory out of materials used to make clay pigeons, the idea being that when they were thrown they burst on impact to leave their tell-tale smear of colour. They were quite a success, the commissioner of the Metropolitan Police ordering 100 for £12 10s and the National Provincial Bank taking 575 for distribution to their various branches.

Harry Lawrence was also experimenting with a form of rocket-propelled clay pigeon fitted with wings. His first trials took place at the bottom of his father's garden with the potting shed as laboratory. When he thought that all was ready, Harry took his invention out into the meadows around Pinner and, together with his brother Ralph, loosed it off and shot at it. He hit it first barrel, whereupon the clay pigeon turned round and chased both of them all round the field. It gave them such a fright that they abandoned the idea from then on.

Enterprising and ingenious though these inventions were, the main gun-making skills were the only hope of keeping the company's financial head above water, and luckily demand for guns slowly improved, including requests for big-game rifles from India and Africa. A double-barrelled .600 bore big-game rifle was made for the Maharajah of Mysore. It weighed 15lb 10oz and gave Bill Morgan such a whack in the shoulder when he tested it at the grounds that he kept the bruises for weeks. A pair of .577 double rifles

A shoot at Glemham Hall, Suffolk, in 1933. Front, from left to right: Captain Ivan Cobbold; his son, John; his daughter, Jean

were built for Mr Raymond Guest at the same time; they weighed 12lb 14oz and both of them added to Bill's bruises but gave him a marvellous new story to relate to his clients.

Further trips to America during 1935 and 1936 brought in enough orders to keep the factory full of work and also to use up a large proportion of the stock guns which had been built in the early thirties and were waiting to be finished off to customers' requirements. The English orders for new guns increased to 226 in 1937. However, the Spanish market, which had always been one of Purdey's best had, to all intents and purposes, ceased to exist after the Prince of the Astorias, later to use the title of the Count of Barcelona, and father of the present King Carlos, ordered a pair of guns in 1927; the Marques of Manzenado, one of the most stylish of shots, bought guns for game and competition pigeon shooting in 1930, as did his half-brother and equally brilliant shot, the Conde de Teba.

Count Teba, incidentally, continued to use these guns until 1971, when they were destroyed in an extraordinary accident. While partridge shooting outside Madrid, the Count, with his loaders and dogs, was standing close to a wall behind which was a stud farm for breeding fighting bulls. The noise of the firing infuriated one of the bulls which managed to scramble over a broken part of the wall and advance on the line of guns. All took to their heels — men, women, loaders, beaters — except for Teba's dogs who went straight for the bull and started to mob it; but the bull turned on the two dogs and looked as if he was having the best of the fight. Teba loved his dogs, so loading one of the guns which the loaders had dropped in their panic, he went to their rescue. The bull saw him and came slowly towards him, head up. Just as Teba was about to fire, the bull lowered his head and charged, and the full shot-charge from both barrels went into the hard gristle between his horns at about fifteen yards range. So strong and massive was the animal that this terrific shock had no effect on him at all; one of his horns caught the Count just above the knee on the inside of his right leg and then flung him high in the air over the bull's back. The dogs rushed up to save their master and drew the bull away, but unfortunately the skirmishing took place right over the spot where the two guns had fallen and the bull's feet smashed them beyond repair. Count Teba was saved by the beaters, made a full recovery, and ordered another pair of guns.

However, in the 1930s the situation in Spain was such that virtually no orders were coming out of the country, and after 1930 only five guns were ordered until 1946. King Alfonso XIII left Spain on 14 April 1931 without abdicating his throne and the political unrest continued until July 1936 when Civil War broke out; after that the prospects of trade virtually ceased to exist. The King, so loved by everyone in Purdey's, died in Rome on 28 February 1941. He had written to Tom from the Savoy Hotel Fontainebleau, on 31 December 1931:

> Happy New Year believe me yours sincerely
> Alfonso R.
> 31. XII. 1931.

Tom was miserable: 'The nicest man and best friend this firm has ever had', he wrote besides the king's name when closing his account in the ledger.

To add to these troubles were the usual complaints from customers, the worst being from a proud owner of a brand new pair of guns who was shooting grouse in Scotland in 1932. On the Twelfth of August he waited in his butt for the grouse to appear. He raised his gun and pulled both triggers: 'click click'. Changing guns he did the same: 'click click'. Furious with rage he telephoned Tom and sent the guns south. On examination it was found that 'slave' plungers, normally only fitted during 'proof', had been fitted to the guns by mistake. One of Tom's famous tantrums ensued: everyone was summoned to the Long Room, and factory manager, finisher and viewer all came in for their share of wrath. 'When a man pays this sort of money for one of my guns, he does at least expect the bloody thing to go bang!' roared Tom.

The strain of all these difficulties of the 1930s told on Ernest Lawrence as factory manager, and in 1934 Tom and Athol decided he needed a rest. So having paid for a cabin for him and his wife in the *Queen Mary*, and given them £50 spending money, they sent them both off to New York for a well-earned holiday. They stayed for a week at Mayfair House, a very pleasant hotel, and returned to England after a whirlwind week during which they toured New York and went to the top of the Empire State building, at that time the highest building in the world. On his return, Lawrence's salary was raised from £7 to £10 a week.

Other customers of the 1930s included the German Ambassador to London, von Ribbentrop, who was disliked by the front shop staff for

(Overleaf) The Long Room in 1934

coming into Purdey's flanked by two burly guards. Prince Frederick of Prussia, grandson of the Kaiser, struck up a great friendship with Tom. After war was declared in 1939 the prince was interned and worked on roads in Northumberland. He used the English name of Mr Mansfield and astonished Eugene Warner in 1944 by walking into the shop to visit Tom. 'I've been granted a bit of leave' he said.

Rear Admiral A. Dudley Pound, who was to be First Sea Lord during World War II, bought a game gun, and the King of Rumania and Queen Marie of Yugoslavia bought pairs of guns. The Maharajah Jam Sahib of Nawenager who shot from the left shoulder was unfortunately blinded in the left eye by a golf ball, so sent in nine single and double rifles and twelve shotguns to be restocked with 'cross-eyed' stocks in order that he could shoot from his left shoulder but use his right eye, all the cases for these weapons having to be altered to take the new measurements. Five thousand cartridges were loaded and sent to the British Ambassador in Kabul.

In 1935, changes were again taking place in the board room. The Oliver brothers decided to sell their holding of shares, and resigned from the board on 30 May. The new owners of the shares were Lord Portal of Laverstock and Major Godfrey Miller-Mundy whose home, Redrice in Hampshire, had a famous pheasant shoot. Loans were made to the company by both these gentlemen, and Tom and Jim were voted £200 each in view of their special services. As Tillson, the secretary, was now seventy-seven, an assistant secretary, Ted Howes, was appointed. At a later board meeting the same year it was decided that, as orders were coming in so well, and profits had improved, it would be possible to pay the £2,500 arrears of dividend on the preferential shares from 1933 onwards, and an extraordinary general meeting was held to sanction this proposal. In addition Tom and Jim, to protect the goodwill and safeguard the firm, covenanted not to carry on business in competition with the company for a period of ten years and, in return for doing so, cheques were drawn in their favour for £5,000 and £4,500 respectively. No dividend was recommended for 1935, and new auditors were appointed.

On Monday, 20 January 1936, King George V died at Sandringham House. He had been a customer for so many years, and was so loved and respected by the Purdey family of three generations, that his death was felt as a personal loss. In the ledger beside the King's name Tom wrote: 'The kindest and best customer we ever had, Tom Purdey, Tuesday 21st, 1936.' The King had been consistently appreciative and grateful for everything they had tried to do for him, and as a result every effort had been made to give him the best possible service.

Later in the year a book, *The Shot Gun*, by T. D. S. Purdey and Captain J. A. Purdey (Tom and Jim), was published by Philip Allan in The Sportsman's Library. It is a fascinating work on various forms of shooting, both of game birds and big game. It was not intended to give the technicalities of gunmaking, but simply to be a guide for young boys as to the best way to enjoy their sport. In it they quote the late King:

> When you are shooting badly the following advice given to one of the writers by our late King may help: 'Stop shooting — smoke a cigarette and watch their flight; then when you start to shoot again don't look for the birds until they are nearly over you — then shoot.'

They also give advice to customers about gunmakers.

> Make your gunmaker your friend, and don't look on him as a robber. If the pence are scarce tell him so, and he will always meet you as regards his charges; but don't do what some people do — send their guns in and have all sorts of work done and then first say they didn't authorise the work, secondly complain of the prices charged, and finally refuse to pay. It never impresses the gunmaker and cannot be much satisfaction to themselves. Fortunately this type is not numerous. Be kind to your gunmaker and you'll get the best service and help he can possibly give you.

Sir Wyndham Portal and Tom in Monte Carlo in 1935

Written in 1935, this holds good today as it held good then, or indeed in 1814 when the first James Purdey opened the doors of his shop off Leicester Square.

The Eastcote land had been sold and Tom had difficulty in finding another field suitable for shooting grounds. In October 1936, however, he was fortunate in arranging with Mr Richmond Watson for Purdey's to use his facilities at the West London Shooting Grounds at Northolt for an experimental period of one year at a generously low rental, during which the firm could test its guns and rifles and eventually install its own instructors. This arrangement worked excellently and, thanks to the generosity and kindness of the Richmond Watson family, has continued ever since. After the retirement of Bill Morgan on 31 December 1954 the instructors at the grounds, notably Percy Stanbury and Michael Rose, took on Purdey customers and trained them to shoot.

On 11 December 1936 King Edward VIII abdicated; and on 12 May 1937 King George VI was crowned in Westminster Abbey. Audley House was decorated for the occasion with strings of red, white and blue bunting to the top of the building and across every balcony. Blue boards were fitted to the railings of the first-floor flat and painted with 'God Save the King' in gold. Huge Union Jacks were hoisted on the two flagpoles, and the crystal decoration was erected over the door with 'G VI R' on the central panel. The Coat of Arms was painted and Tom gave a series of dinners in the Long Room. He himself was invited to see the Coronation Review of the Fleet from the aircraft carrier *Glorious* commanded by his old friend of Mediterranean adventure days — now Captain Bruce Fraser. In keeping with this mood of jubilation, trade had improved satisfactorily since the Depression and the net profit for 1937 reached £8,500. The orders for new guns amounted to 226 from all countries in that year, but shadows of war were already lengthening and Tom was wondering about the future of the firm should conflict once again break out.

Meanwhile Lord Portal and Major Miller-Mundy, like their predecessors, insisted on economies on the part of Tom and Jim. As their generosity had saved the firm from bankruptcy, they were naturally incensed when they discovered that Tom was carrying on entertaining his friends in the usual lavish way and charging most of it to their business. He was summoned to Laverstock and given a good dressing-down by Lord Portal for his duplicity. Very contrite, poor Tom retired to bed; but Lord Portal was furious to discover after his departure next morning that he had sneaked downstairs during the night and helped himself to a liberal nightcap from the decanters in the dining-room. Tom was immediately summoned again, but said that he had no car to take him down to the country. Wyndham Portal said that he had no intention of providing him with a car after what had happened, and Tom was to buy a bicycle, put it down to the firm, and ride on it to Hungerford. Tom got round this by ordering Graham Tollett to ride the bicycle down while he travelled by train and taxi. Wretched Graham took ten hours to reach the house, was exhausted when he arrived, was given a quick meal and then returned to London by train. Tom's behaviour in the factory was even more autocratic: 'Keep your head down and get on with your work', was his usual form of greeting when he did his tours.

In December 1937, Jim was sent to America on another selling trip. He took thirty-two orders but this was against forty the year before. The following year, 1938, American orders fell to fifteen. In January 1938, the new King gave a pair of guns to King Farouk of Egypt. They were presented in Cairo by the British Ambassador, Sir Miles Lampson. In May, Tom told the board that a scheme had been drawn up under which all workmen would be granted a week's holiday each year with pay. This was fully approved by the board.

During February the Long Room, which had been in urgent need of repair for some time, was rebuilt. The central well, through which the Purdeys were able to watch the work being undertaken by the storage staff in the gunroom below, was filled in. A table was put down the centre of the room, and an improvement was made at the north end to the staircase which gave access to the basement rooms. Tom, a keen musician, installed two pianos, a harmonium and a large sofa. Dr Malcolm Sargent and the future Earl Alexander of Tunis would play duets with Tom, and customers were actively discouraged from entering, as both Tom and Jim considered this room very much their personal drawing-room.

But by now, Germany was advancing over Europe. Athol, now in his eighty-first year, had to watch the same pattern of events taking place as before World War I — all markets collapsing for his family firm, the switchover of skills to the standardisation of war work, and the danger of young men having to leave their benches to go and fight. The difference on this occasion was that Tom would now have the responsibility for holding the firm together during the coming battles, and although Ernest Lawrence was still factory manager, his son Harry was to have the duty of converting the factory to the requirements of the war effort. History had repeated itself with extraordinary precision but with one important variation — whereas France, Spain, India and the United States still continued to be supplied with shotguns and rifles for two years after the outbreak of war in 1914, on this occasion all markets ceased to exist. Few Purdey guns were to be built for six years.

Knowing all too well what was coming, Tom told the board in September 1938 that he had written to the Air Ministry regarding the possibility of carrying out special work for the Royal Air Force. Orders for guns had now virtually ceased and there would be no work for the men after the next

Three generations of Purdeys, in 1938. From left to right: Jok, with his grandfather Athol; his father, Jim; his stepmother, Mary

nine months. In order to save the £3,000 considered necessary to prevent the firm from going bankrupt, it was proposed to lay off fifteen of the craftsmen as contracts from the ministry seemed a remote possibility.

But at this point the efforts of Harry Lawrence to find suitable alternative work for the workforce bore fruit. A firm called Aircraft Materials near King's Cross station wanted someone to design and make a 'pop rivet' tool for 'stress-skin air frames' for the Air Ministry. Harry made up one of these tools in the factory and presented it for trial. The firm was so pleased with the results that it placed a large order which was worked on by the men at the factory in Praed Street. In the meantime, Tom Purdey had been told by BSA in Birmingham that the manufacture of the Short Lee-Enfield Mark IV rifle was being speeded up as war was not only expected but imminent. There were not, however, sufficient skilled gauge-makers to make the necessary gauges. Harry Lawrence was sent to Birmingham to find out the true position, and as a result of his conversation with the BSA directors, a contract was given to Purdey's for this prewar work.

While the firm was altering all its priorities to meet the threat of war, the last of the usual orders for repairs, cartridges and guns were being completed; and on 18 May 1938 an eminent physician, Sir Thomas Dunhill, was supplied with quantities of sizes 5, 6 and 8 lead shot. He had a patient who was unable to swallow, and as a consequence had to be fed through his side. After operating, Sir Thomas found that the patient was unable to make full use of his throat muscles, so in order to stimulate these into proper use, he threaded the shot onto silk thread and lowered the 'beads' down his throat.

Tom's personal appreciation of his customers had become more controlled over the years. He wrote against an entry of 1938 for Captain Robin Grosvenor, the future Duke of Westminster: 'Very Special. Will be D of W and our Landlord. T.P.' And as the firm had not done badly during 1938, the board voted a capital sum of £500 to be distributed between Athol, Tom and Jim.

On 30 April 1939 Athol Purdey died at Folkestone after a short illness aged eighty-one. He was cremated at Ashford Crematorium and his ashes were interred in its grounds on 3 May, a memorial service being held at St George's, Hanover Square, the following day. The obituaries in the *Shooting Times and British Sportsman*, the *Field* and many other newspapers referred to his geniality and to his great knowledge of guns and the great shots of the past, all of whom he knew personally and had advised. He had been born during the Indian Mutiny and had served Queen Victoria, King Edward VII, King George V, King Edward VIII and King George VI in a personal capacity, holding the Warrants of Appointments and the personal regard of almost every crowned head of Europe. His contribution to the continuity of his family business throughout World War I and a major depression was complemented, as in his father's case, only by his service to the Gunmakers' Company and the maintenance of standards of safety to the public in matters of gun proof. He died as yet another war was threatening the future of his family and his firm. Mabel living in his flat above Purdey's, Tom and Jim who were running the firm, the Lawrence family and Eugene Warner who loved him dearly as did all the men in the factory and who mourned him with great respect, never guessed that he would die intestate. At a board meeting held on 15 August 1939, Tom reported that the majority of the furniture, pictures etc in the offices at South Audley Street belonged to the late Athol Purdey and he thought it desirable that these should be acquired by the company.

On 5 May 1939, Sir Harry Stonor, the last of the great shots died in St James's Palace aged seventy-nine. On 20 January that year he had written to Athol Purdey telling him that he had decided to give up shooting:

My dear Purdey,

    Ever so many thanks for your letter with the enclosed cheque for my old guns; I am so sorry to part with them but I cannot suppose I shall ever be able to shoot again. I am off to the Hotel de Paris at Monte Carlo on Sunday and shall hope to see you out there.

                        Yours ever
                            Harry Stonor.

Mabel Purdey died in her Audley Street flat on 5 September 1939. She left no will and few possessions. Her sister-in-law, Constance Svedberg, had been giving her an allowance of £50 per year. She was cremated and her ashes were buried with Athol's at the crematorium in Ashford.

WORLD WAR AGAIN

World War II had arrived and once again the normal life of everyone in Purdey's was to be completely disrupted. The organisation of the factory had to be changed to deal with the requirements of the various ministries who were desperate for skilled workmen to carry out their designs for precision tools; but in this war the threat of devastating aerial attacks was present as well, so entirely novel precautions had to be taken to comply with the 'black-out' regulations. The skylights in the Long Room were painted dark blue and all work had to be done by artificial light. Dark blinds were hung over every window of Audley House and the factory, and each pane of glass was criss-crossed with bands of sticky paper to stop glass splintering into the faces and eyes of the occupants of the rooms. Gas masks were kept ready in their little cardboard boxes, and tin hats for the men who would act as fire watchers on the roofs. Buckets filled with sand, and stirrup pumps with buckets of water, were placed in convenient positions to deal with any fires which might be started by incendiary bombs.

Some of the more valuable contracts hoped for by Tom and the board had not materialised, but the Ministry of War had thousands of the P14 .303 rifles, the standard rifle of World War I, safely stored away, heavily greased and wrapped, and all these needed degreasing and preparing for use. Purdey's was one of the firms approached to do this work and the boxes of rifles were delivered from the Ordnance Depot in Woolwich to the factory at Praed Street. The craftsmen were taken off their various gunmaking tasks and divided into teams of six. The grease plastered thickly on the bolts, actions, barrels and sights of these old rifles had been there for twenty years and it was a filthy job. But the factory worked steadily through this huge pile of work, completely renovating rifle after rifle, each of which was then inspected and approved by a team of inspectors from the Royal Small Arms factory at Enfield, and the firm received 10s for each accepted weapon.

One or two of the older men such as Charles Lane a finisher, and 'Baldy' Williams who could not adapt to a new way of working, were kept on as gunmakers to complete the guns still on order.

Tom had been ill for most of winter 1939 and spring of 1940, so the reorganisation was conducted by Jim and the two Lawrences, father and son, who had finished the conversion of the factory by April of that year. Not only had they started the new war work programme, but they had also completed the orders for shotguns and rifles taken in 1938 and 1939 and those guns, built for foreign customers, were stored at Audley House for delivery when the war was over.

In the autumn of 1940 the German airforce started to bomb London, and at a board meeting held in the Long Room on Thursday, 28 November, 'during an air-raid with enemy planes overhead', it was reported that the Irongate Wharf factory had been hit by incendiary bombs and the roof damaged by falling debris from a nearby building. This damage had been repaired by the men in the factory; two windows in South Audley Street had also been blown out by blast. As the Ministry of Defence needed more and more rifle and bullet gauges extra premises had to be found, and Harry Lawrence managed to acquire part of the premises of Reo Motors on the Great West Road at Brentford, and two-thirds of the skilled men from Praed Street were transferred to this new factory. The Ministry of Supply provided the machinery on the total value of which the company was charged 12½ per cent per annum for a period of eight years. Fortunately quite a few of the men lived in the Brentford area so the difficulty of getting to work in wartime London — which when few workers owned cars and there was strict petrol rationing in force depended on buses, undergrounds, bicycles or walking — was made easier. In fact the only man in Purdey's apart from Tom and Jim to own a car was Harry Lawrence, whose little Armstrong Siddeley was used to help get men to work and for general factory deliveries.

The work carried out in the new factory consisted of making gauges for BSA in Birmingham who were building the Mark IV rifle, gauges for Rotax of the aircraft industry, bullet gauges for Imperial Chemical Industries (ICI), gauges for the building of the 9mm sub-machine gun and plastic moulds for the Air Ministry. The number of skilled men at Purdey's, fifty-seven, was insufficient for these activities so sixteen extra men with engineering skills had to be recruited and it was then, with this influx of engineers, that a union — the Amalgamated Engineering — came into the firm. Men could join if they wished to do so, many of the gunmakers did, and a shop steward appeared on the scene. This new approach to life was completely alien to anything the two Purdey brothers had ever known, and just as difficult for the Lawrences to comprehend; but they just had to get on with it and, mainly due to tactful handling by Harry Lawrence, gunmakers and engineers were able to work well together.

On the night of 16-17 April 1941 there was a particularly vicious air-raid on London, and at ône o'clock in the morning a stick of four bombs hit

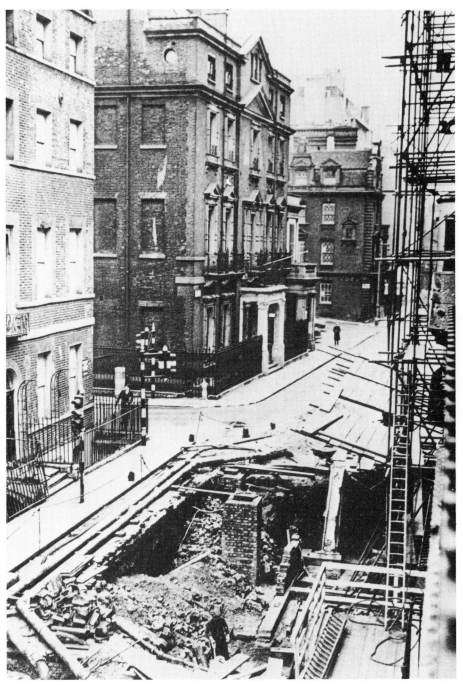
South Audley Street after the bombing on 16 April 1941

South Audley Street and Mount Street. Every window was broken, the slates were blown off the roof leaving the cartridge shop open to the sky, and all the original roof-top decorations were destroyed. Tom Purdey and his Staffordshire terrier Johnny had been in the third-floor flat when the bomb fell, and were found next morning sitting in the downstairs kitchen in the company of Tom's manservant, William, who lived in the basement, covered in plaster and in a state of shock. The lovely railings originally installed by his grandfather and surmounted by Royal Crowns had been blown away, and a huge piece of shrapnel was embedded in the wall at the bottom of the staircase. The pillars on the outside of the building were scarred and cracked by shrapnel and remain so to this day, an honourable and permanent reminder of the bombing of London. Old Mr Lawrence gathered the men together from the factory at Irongate; they scaled to the top of the building with tarpaulins and covered as much as possible. They also boarded up the windows until the glaziers could get round to replacing them. The Long Room, surrounded by buildings, was untouched.

With the whole emphasis on war production, night shifts were introduced, the men taking cover under their benches whenever the bombs seemed to be getting too close for comfort. A difficulty had developed in the manufacture of the chamber caps of the new 9mm rifle, so another department had to be brought into existence to deal with this. A warehouse was rented next door to the Paddington factory and twenty-five women were employed under the supervision of Christopher Gadsby, then a finisher and now factory manager, to make the special tools and jigs for this operation. Harry Lawrence invented the tool that eventually corrected this fault, and in 1943 that part of the operation was in full production.

The men in the factory, however, were not only producing all they could during the day, but at night carried out extra duties such as that of air-raid warden and, in the case of Chris Gadsby and Dan O'Brien, served on the rocket sites in Hyde Park. Gadsby rose to the rank of sergeant in a rocket battery situated close to Park Lane, and after one particularly awful air-raid woke next day at home to find that all his hair had fallen out and was lying on the pillow beside him.

Lord Portal of Laverstock and Mr Harold Macmillan were Joint Parliamentary Secretaries to the Minister of Supply, Lord Beaverbrook, and as close friends of the Purdey family they were invited to a party to celebrate Jim Purdey's fiftieth birthday which was held in the Connaught Rooms on 30 September 1941. To mark the occasion a portrait of Jim had been commissioned from H.M. Broadhead, every man in the firm being asked to contribute to the cost so that the firm could present the picture to Jim personally. Each man gave between 2s 6d and 5s and the entire firm took the day off for the event. Eighty-six men and twenty-eight guests including Jim's wife Mary; Sholto Douglas, Tom's old commanding officer in 43

Squadron, now Air Officer Commanding Coastal Command; Harold Balfour, who had been Tom's companion in the same squadron during World War I and was now Under Secretary of State for Air; Sir Humphrey de Trafford; Colonel Ivan Cobbold; Lieutenant Colonel Tim Nugent, later Lord Nugent; Stanley Harley, expert on steel moulds and a director of Coventry Tools Ltd; the American actor, Robert Montgomery, then an assistant naval attaché in the American Embassy; General Sir Adrian Carton de Wiart VC; Sir Stewart Menzies, the head of MI 5; Cecil Purdey and Jim's son, Jok, now a lieutenant in the Royal Welsh Fusiliers, sat down to a meal of:

Green Pea Soup

Roast Turkey and Sausages
Vegetables

Fruit Salad
Icecream, Wafers

Coffee

For wartime England this was a great treat and a well-earned day of relaxation and fun, as the men had been working seven days a week, night and day shifts, with no holidays since June 1940. Harold Macmillan proposed the health of the firm, Jim was presented with his portrait by the artist, and when one old man was asked what he thought of it he replied that it looked like a blindman reading Braille. After the presentation and the loyal toast, everyone present was issued with a sheet of paper on which was typed a very patriotic 'Victory Song' written by Tom to the theme of Beethoven's Fifth Symphony. The RAF Salon Orchestra had been playing throughout the meal and they now played for the office staff to sing Tom's composition, and as they reached the words 'Three dots a dash' followed by 'This is the sign of Victory' everyone had to bang their fists on the table as loudly as possible:

> Hear now a song of the Empire.
>   The song of the men of the air,
> This is the song of the men of the seas,
>   And the soldiers beyond compare.
>
> This is the song of the workers,
>   Louder and louder it grows,
> Caught on the breeze, flung o'er the seas,
>   Till this mighty chorus flows.

> Three dots a dash, three dots a dash.
> > This is the sign of Victory,
> Three dots a dash, three dots a dash,
> > This is the song of the V.
> Free men and men in chains,
> > Where so'er they may be,
> Sing, mark and pass the word
> > Victory and peace for the free,
> Three dots a dash, three dots a dash,
> > This our anthem shall be,
> Let us unite, with all our strength,
> > And pray for Victory,
> Send him victorious, long to reign over us,
> > Lift up your voice, lift up your hearts,
> And pray for Victory.

Although this song had no great merit, Tom was particularly proud of his composition. The sight of the distinguished guests and all the men solemnly singing and banging with their fists was hilariously funny!

Everyone was enjoying himself enormously before returning to the factories for the night shift when the secretary, Ted Howes, was called away to the telephone. A reporter was on the line and next day the *Daily Mirror* carried banner headlines and write-up on its front page:

> Happy Birthday for you, Capt. James Purdey.
>
> A large London factory, engaged on important war work closed down for the day yesterday. It held a party . . . Mr Harold Macmillan, Parliamentary Secretary to the Supply Ministry gave a 'Buck-up speech' during the banquet. Maybe this made everyone feel all right . . .

On 7 December 1941 the Japanese attacked Pearl Harbor and the United States entered the war. American servicemen started to arrive in England during January 1942 and Tom and Jim, with their friendships and contacts in America, became hosts to a great number of their customers who came to England in the United States forces. The Long Room became a meeting place for their American friends; and General Bedell-Smith, General Eisenhower's Deputy Chief of Staff, used the room for conferences leading up to the planning and execution of the D-Day landings on the coast of France. The front-shop staff were particularly proud on the occasions when General Eisenhower himself came to Purdey's for these meetings.

Cartridges now were in great demand, not for game shooting, but for the control of rabbits and vermin. As the vast majority of gamekeepers had joined up, farmers were inundated with unwanted creatures of all kinds on their land, and the cartridge shop over South Audley Street was working full-time to meet landowners' requirements for their tenants. Very little shooting had taken place during the season 1939–40, and the winter had been exceptionally hard. There were, therefore, large numbers of hungry pheasants preying on crops. As a result of a suggestion made in the *Daily Telegraph*, an Order in Council was made extending the date of pheasant shooting by one month to 1 March in order to protect the farmers' interests. During 1940, 1941 and 1942 some 922,000, 570,000 and 611,000 cartridges respectively were loaded by hand in crude brown-paper cases for customers and the stocks built up before the war became exhausted. Both cases and wads were of very inferior quality but they were the best material which could be obtained, and in spite of the bombing in April 1941 the loading shop under Ralph Lawrence maintained this pressure.

As a result of the war work contracts, the net profits of the firm rose dramatically as they had done between 1914 and 1918. After a drop to £4,700 in 1940, an average figure of £8,000 per annum was reached and maintained until 1945, mainly after a bonus system was introduced by the firm in the autumn of 1941.

In the scrapbooks kept by Tollett, who each morning had to go through the papers and glossy magazines to cut out references to customers and the firm, are photographs of Tom's friends and acquaintances in uniforms of every conceivable type. Field Marshal Mannerheim is there wearing a big white furry hat above his bemedalled uniform as well as Robert Montgomery as a lieutenant in the US Navy; Lord Louis Mountbatten as a Royal Navy captain; the cricketer, George Newman of Middlesex and Oxford, as a sub-lieutenant, RNVR; David Niven in the uniform of the Rifle Brigade; Lord Marchwood as a Home Guardsman; Mr N. S. McCorquodale, Member of Parliament for Sowerby in Yorkshire, in the overalls and tin hat of an air-raid warden, ringing the 'All Clear' on a dinner bell; and members of the Royal Family in the uniforms of all the services. There is also every cutting possible concerning the exploits of the Argyll and Sutherland Highlanders, Tom's original regiment in 1915, whose gallantry in the fighting in Malaya before the fall of Singapore gave him immense satisfaction and pride.

On 25 August 1942, the Duke of Kent was killed when the Sunderland flying boat in which he was travelling to Iceland on the staff of the Inspector General of the RAF crashed in the north of Scotland. Tom wrote to the Duke's mother, Queen Mary, and received a charming letter from Her Majesty's lady in waiting saying that the Queen 'is deeply touched by your kind thought in writing having known you and your brother so well when you were quite boys'. The Duke of Windsor also wrote from Government House, Nassau, telling Tom of his great grief at the loss of his younger brother.

On 23 October, General Montgomery won the Battle of Alamein, and

MARLBOROUGH HOUSE
S.W.1.

5 Sep. 1942.

Dear Mr Purdy

I am commanded by Queen Mary to thank you & Mr Tom Purdy warmly for your very kind letter of condolence in the fearful calamity which has so suddenly befallen Her Majesty. The shock has been a very terrible one, and her sense of her parting is indeed hard to bear, but the wonderful messages of sympathy

which have been showered on the Queen from all over the world have been a very real comfort to Her Majesty — who is deeply touched by your kind thought in writing, having known for some years so well when Princess Elizabeth.

As you say, it is a consolation to the Queen to think that the Duke died on active service, + that he had (in deed) a relinquished of his rank (of ?Marshal) for a humbler & Air Marshal for a humble + more active position in the R.A.F.

Yours very truly
William ?

the tide of war turned. In the factory, work continued at the same pace as before and with the same pressure for delivery from the various war departments. Already there were doubts in the minds of the directors as to the feasibility of returning to the manufacture of guns after the war was over, as taxation was so high and the men were gaining such a good name for the firm with their work on tools and moulds that it looked as if this sort of work would become more profitable than guns. Cecil Purdey brought up the question at a board meeting in May 1943, and Tom advised the members to keep an open mind about the matter; he also pointed out that it was most important to watch the size of the wages cheque in relation to the figure of net sales. Pressure from the shop floor was beginning to squeeze profit margins.

But all was not well with the Purdeys themselves. The strain of war, the bombing, and their precarious financial position was taking its toll. Lord Portal wanted to build up the firm's financial resources, but Tom and Jim needed money; and so bad did the position become that Jim wrote secretly to Wyndham Portal on 30 April 1943, telling him that Tom was on the verge of a nervous breakdown caused mainly by his financial worries and debts. Jim had seen the yearly accounts and, while he understood that the future outlook was obscure for all of them, added 'Could you save the situation and Tom's health by a more generous distribution than that supplied?' After paying the preference dividend and making provision for reserves and taxation there still remained £4,000 which could be distributed to the shareholders, and Lord Portal had only allocated £200 each to Tom and Jim for expenses. Tom by now had taken the matter into his own hands and had written a furious letter to Mr Bell the accountant. The latter and Lord Portal were also war weary and the inevitable row erupted. Bad feeling started to creep in and Lord Portal felt that he had had enough. Other financial backing would have to be found.

In the meantime, Tom's state of nerves had embroiled him in a difficult situation in the factory. Jim was living at Wentworth, so Tom used to travel down to stay with him and Mary most weekends. After a particularly good lunch at his club he decided to break his journey at the Great West Road factory. While walking round with Harry Lawrence some altercation took place and Tom, in a loud voice, called Harry 'a bloody fool' in the hearing of several of the men, who, after Tom had gone on his way, proposed a strike unless Tom apologised to Harry in public. So indignant did the men become that Harry's father felt action should be taken as soon as possible to prevent a nasty incident. A deputation from the factory therefore presented itself in the Long Room, young Ben Delay, now the senior actioner,

The letter sent from Queen Mary to Tom on the death of the Duke of Kent in active service

Tom being taught to fly a glider in 1943. He was with Harold Balfour, who was then Joint Under Secretary of State for Air

being one of the members. Tom handed out drinks and the matter was discussed at length, Tom apologising for his behaviour and everyone agreeing that the matter was now closed. Jim Purdey, on the other hand, put the whole matter in the correct perspective. 'This is balls', he announced to the deputation, 'Mountains and mountains of balls', and when everyone had left Tom told Harry that he could not understand what all the fuss had been about, as he had been calling him a bloody fool for years and no one had ever complained about it before.

In July, Lawrence Salter, nephew of Harry Lawrence and grandson of Ernest Lawrence, joined the firm as a gunmaker/toolmaker in the Great West Road factory. There were no apprentices as such at that time but, as in the case of so many generations of beginners before him, his first job as a trainee was to find out what the older men wanted for their lunches and then run round to the shops and buy it for them, being careful not to mix up the orders and bring back the correct change. After that the future managing director of the gun firm started learning the basics of engineering. The effect of a union shop was being felt and Tom Purdey, who had given £5 as a prize to be played for by the darts team, almost had an apoplexy when he received a letter of thanks from the self-styled shop steward which read: 'Dear Comrade Purdey, Thank you so much for the money.'

WORLD WAR AGAIN

In November 1943 both Lord Portal and Major Miller-Mundy sold their preference shares in Purdey's to the firm of Cobbold and Co Ltd of Lower Brook Street, Ipswich; and Captain Ivan Cobbold became owner and director of the firm at the same time. He had been an old friend and customer of the Purdeys for many years and was one of the best shots and keenest sportsmen of his generation. At this time he was a temporary lieutenant colonel in the Scots Guards and held a responsible position in liaising with the United States forces. The strength of purpose and financial backing which he brought to the Purdeys was a great support, and Tom and Jim felt the future to be in very good hands, with their company in the charge of a man of such vitality and intelligence. The first improvements he made were to increase the salaries of the responsible managers of the firm and the secretary, and to give bonuses for war work to the front-shop and clerical staff; this inevitably increased confidence all round. He also gave a special bonus of £250 each to Tom and Jim for the strain under which they had worked since 1940. On 30 October 1943, Cecil Purdey died in hospital after an operation. He was the last family link in the firm with his father, James the Younger.

On Boxing Day 1943 the German battleship *Scharnhorst* was caught and sunk by ships of the Royal Navy commanded by Admiral Sir Bruce Fraser, Commander-in-Chief Home Fleet, flying his flag in the battleship *Duke of York*. Tom wrote to congratulate his old friend of *Resolution* days of 1920, and received the reply shown overleaf. After this wonderful news had been duly celebrated in the Long Room, 1944 opened a new era for Purdey's.

War work continued, but a new mood was alive in the hitherto very happy factories. The main contention was money. The scales of pay were laid down in accordance with the profit levels agreed between the firm and the various ministries who supplied the contracts. After discussion between the firm's accountants and the ministry representatives, agreement was reached on the rates of overheads on labour, materials etc; the Ministry of Supply then allowed the firm a small profit. There was not, therefore, much room for manoeuvring on wages, and this was pointed out to the men by Harry Lawrence. Nevertheless the constant nagging at management continued, making Harry's life a misery.

The strain and worry of wartime conditions was increased on 6 June by events connected with the invasion of Europe, followed seven days later by the first flying-bomb ('buzz-bomb') attacks on London. The psychological effect of this weapon was intense, and the firm was further shattered by the news that on Sunday, 18 June, Colonel Ivan Cobbold had been in the Guards Chapel when a flying-bomb scored a direct hit on the building. The colonel was killed instantly and Tom, who had planned to be with him, was only saved from the same fate because he had overslept and could not get down to Birdcage Walk in time for the service. General Bedell-Smith

Commander-in-Chief,
Home Fleet,
c/o G.P.O., London.

4th January, 1944.

My dear Tom (and Jim)

Thank you so much for your ever welcome letter of congratulations. It was so kind of you to write. It was indeed a very great moment for us all and we are glad that we have been lucky enough to be able to give you a small Christmas present.

*Yours ever,
Bruce Fraser*

Tom Purdey, Esq.,
 Audley House,
  57, South Audley Street,
   London, W.1.

had also been invited by Ivan to attend; he was dressed and ready to go when an important message demanding his immediate attention arrived from Washington, so he stayed in his office. This dreadful tragedy was felt very keenly by Tom. Ivan Cobbold had been his best friend and had supported him and the firm in many difficult situations. Not only had his wise guidance been removed but also his close friendship, and Tom was a man who relied for confidence on the support of his friends. It was a very dark time for the firm, especially as the guiding hand had been removed just when firmness and tact was needed to deal with money troubles in the work force. A further worry was the cancellation of contracts on which the firm was relying when, during August 1944, the ministries involved foresaw that the end of the war was near. One of the factories in Praed Street had to be closed as a result, several of the women workers being made redundant. Although this saved the firm the costs of their wages and the rentals of the machinery loaned by the Ministry of Supply, the action only added fuel to the trouble among the men.

The orders for guns during the war had diminished steadily. While guns ordered before 1938 were completed by the few old craftsmen available, customers still placed new orders on the understanding that no guns could be built until the war was over, when their guns would be started in sequence of ordering as soon as possible. They were quoted a price of £150-£155 per gun and no deposits were asked for. The orders received during the war years were as follows:

	Total	UK	USA	India	Portugal	Spain	Australia
1941	61	44	10	1	4	–	2
1942	22	10	2	3	4	3	–
1943	32	20	–	5	4	3	–
1944	21	6	1	1	11	–	2

If, therefore, it was going to be possible to start making guns once more at the end of the war, the firm already had 136 gun orders as a base for putting the factory back to normal working.

On 8 September 1944 the Germans added rockets to their armoury for bombarding England, the first one falling in Chiswick, and many people regarded this weapon as even more terrifying than the buzz-bomb as there was no warning at all before a shattering explosion shook the whole surrounding district. The buzz-bombs were still falling however, and Tom went down to the Great West Road factory to boost morale and talk to the shop stewards. In the middle of the discussions the roof lookout reported an approaching bomb. Everyone took cover under his bench except Tom, who announced that he had been in the battle of Ypres and had no intention of lying down because some 'bloody German bomb was coming over'. He stamped about the factory with poor Harry, who felt obliged to stand up as his chairman was with him, at his heels. The men were very indignant in case anything should happen to Harry — they didn't care a fig for Tom provided Harry was safe! On 27 March 1945 the last rocket fell at Orpington, and *The Times* newspaper announced that there had been 146,760 civilian casualties in the United Kingdom due to enemy action since the outbreak of war.

At the beginning of April 1945 the last of the contracts for the Air Ministry came to an end and the 'popper tool' production line was stopped, eleven workers being made redundant. Immediately the shop stewards at the Great West Road factory staged a short unofficial strike and sent Tom a list of questions, demands and proposals. Two of the proposals were dealt with by Tom personally:

Post-war Plans   Post-war plans to be put before the men, in the rough and the final plans before acceptance.
Chairman's Recommendation   Certainly not.

Merit Awards   That employees should be in the position to know how the management sums up the merit of men.
Chairman's Recommendation   By efficiency — as recommended by work and the factory management.

The other proposals were dealt with by negotiation. A second, similar stoppage in June was dealt with by Jim, and at a meeting between Tom and the union shop stewards it was agreed that the latter should be fully recognised and give valuable advice on trade union procedure.

On 7 May 1945 the German armed forces surrendered unconditionally, and the need for war work came to an end. Tom and Jim found their firm in the position of having trained gunmakers making tools who wanted to get back to gunmaking, and an equal number of engineers who could only make tools. Tools were, of course, in great demand for the rebuilding of British industry, but the natural inclination was to return the firm to the trade it knew best. The lease of the Reo factory on the Great West Road was due to expire in January 1946, and it was hoped to find other premises close to the existing factory at Irongate Wharf so that the workforce could be divided into two, one section consisting of the original gunmakers so that guns could be built once more, the other with machines manned by the engineers who would continue to make steel moulds and precision instruments. It was felt that the price of the Purdey gun was high enough at £155, and that the only way to keep the price down was to subsidise the guns through the engineering side of the business. Harry Lawrence had built up excellent working relationships with the rest of the toolmaking industry and hoped the firm could provide them with the precision instruments and moulds they required.

So it was that Purdey's started to build guns once more; but before this could be done the whole gunmaking organisation had to be rebuilt. The gunmakers were settled in the factory in Irongate Wharf and those young men who were thought suitable from the engineering side were brought across to learn their new trade; Lawrence Salter among them.

In November 1945 Harry Lawrence, having assembled as many gunmakers as possible, settled their basic flat rate at 3s 6d per hour with a bonus consisting of the difference between their wages thus earned and the value of the work they produced each week at the 1939 piece rate, plus 50 per cent. As a result of this wage settlement prices had to be increased to a basic £230 per gun, £50 for a pair of barrels and £240 for a pigeon gun. All the customers who had placed tentative orders during the war had to be written to and informed of the new price. The Portuguese firm of Messrs Dias, de Costa, das Corto Silvena and Sousa in Lisbon got a special letter in view of the large number of guns they had on order, and the board were to be kept informed of the replies received.

The toolmakers were now settled in a factory next door to the gunmakers in Irongate Wharf, and by February 1946 Harry Lawrence and his father had done their reorganisation so well that a cash balance of £5,000 could be maintained until it was possible to get guns in production again by approximately the beginning of July.

The award of one MBE (Member of the British Empire) and one BEM (British Empire Medal) was made to the firm for its contribution to the war effort. It was Tom's responsibility to make the allocation and he told his brother Jim that he would love to have the MBE for himself, but 'I can't take it, they must have it'. And so the MBE went to Harry Lawrence for all the wonderful work he had done, and the BEM to Frederick Williams, the oldest gun actioner in the firm, who had kept the gun side going throughout the war.

Tom now desperately needed a new financial backer, but no one thought that a firm such as Purdey's, making very expensive guns, could continue in the climate of taxation and socialism which now existed. Tom, very depressed, went into White's Club and told his troubles to Sir Hugh Seely. Sir Hugh had been Under Secretary of State for Air during the war and had just been created Lord Sherwood. He had great faith in the future so he agreed to buy all the Cobbold shares and, on 4 March 1946, he became owner of Purdey's.

MASTERS OF THEIR CRAFT

The history of the men who built Purdey guns is just as important as that of the family who employed them. Nearly all the factory records of wages and hours worked and even of the number of men employed during the lifetime of James 'the Founder' have disappeared, and it is not until the 1880s that we know that James the Younger employed 83 permanent craftsmen and 25 out-workers and day workers. An extract from the wages books of 1863–92, recently discovered, follows:

Date	No of Men	Top	Lower	Monthly Bill
April 1863	41	£9 7s	£1 2s	£438 10s 0d
April 1865	49			
Jan 1866	53			£502 15s 0d
Feb 1867	60			£587 10s 0d
Jan 1870	64			£610 5s 0d
Advertised for workmen in the *Birmingham Daily Post* (3s) April 20 1872				
July 1877	74			£1120 0s 0d
Sept 1892	81			£1600 0s 0d

1863   G. Field (errand boy) 5s per week

We do not know at what age 'the Founder' recruited boys for apprenticeships, but from the time that he started his own business in 1814 to the present day, Purdey's have trained their own craftsmen in the old English guild system whereby a boy is instructed in one aspect of gunmaking by a 'gaffer', or trained craftsman, and on completion of his training takes his place in the factory as a skilled worker in the particular department in which he was trained. As a trained craftsman he puts his initials on his finished work, so that a complete 'history' of the building of each gun is kept in the firm's records.

From the first, it is essential to realise that those employed in the English gun trade in London and Birmingham during Victorian and Edwardian times were not tradesmen working in awful conditions, fighting for their rights against oppressive masters, and having to strike for the bare necessities of life. Nothing was further from the truth. The British gunmaker was a trained craftsman and therefore independent — able to stand on his own feet by virtue of his standards and skills. Should he not like his employer he could leave him and set up in his own right. If he were sacked, his skills ensured that he could readily find work elsewhere, although if his dismissal was due to bad workmanship other top quality gunmakers would naturally be suspicious of taking him on permanently until he had proved himself. Being a skilled craftsman he demanded a high wage, and as a result lived well in comparison to other workers.

As already stated, from the time of James the Founder, the Purdeys trained their apprentices in the old tried English guild system, whereby a boy came to the factory at fourteen years old and was bound for seven years to a gaffer in one of the departments of gunmaking — barrel-making, actioning, stocking or finishing. The apprentice received no wages for the first two years of his training, but if after that trial period the gaffer decided the boy was good enough to continue training he would pay him for the work he did out of his own earnings.

Apart from learning his trade, the boy fetched and carried for his trainer but, if he did well and learnt fast, he soon found himself being trusted to 'rough off' early stages of work for his gaffer to complete. Every craftsman in the factory worked piece work, which meant that he was paid a set sum when he had completed a 'piece' — eg a barrel or action — and after his work had been inspected and passed by the factory manager, this sum being based on a time-scale agreed with the craftsmen of each department. This system did not necessarily penalise the slow worker as against the man who worked fast, for work skimped for the sake of speed would be rejected by the 'viewer' and the employee involved would have to improve it at his own expense before he received his money. And an apprentice who worked well could cut the time his gaffer took to complete a 'piece', with the result that his trainer could all the sooner start another.

Purdey's best quality guns, handmade throughout, cost £65 to £70 around 1880, with a delivery of eighteen months or more from the time of ordering. These guns were made in sections; each part being made by a fully trained man, who made only that one part all his life. The barrels were built first, then the action, the stock was cut to the customer's measurements, the gun was engraved, the mechanism was regulated and, last of all, the gun was finished. Very great care was taken over the checking of the barrels and the balancing and weighing of the gun. The guns then being built were fitted with hammers and were non-ejectors, although later in the 1880s new ejector mechanisms required another specialised department in the factory, and the introduction of hammerless actions had to be allowed for in the rates of pay.

During his apprenticeship the boy made the tools he would be using for the remainder of his working life. Tools for making barrels are entirely different from those employed for making actions and stocks; so craftsmen in each department used, and use, tools of their own design which have evolved over the years. As each craftsman had his own particular way of doing things, the tools he made were slightly different to those of the man next to him, so no ironmonger's shop could supply his needs. The first job, therefore, that any young man just out of his time had to do was to complete the tools which he would use.

In barrel-making the man made gauges for measuring the outside barrel measurements at various points on the barrel, strikers for smoothing barrels and ribs, and lapping rods and cutters for opening up the inside diameters of the bores. The actioners made jigs for the 'strikers', whole series of clamps for holding plates in position, and their own 'smooths' and files. And since the accuracy of fitting used in gunmaking must be exact, with no question of tolerance, he used smoke from a paraffin lamp to control the fitting of one part to another. The stockers made their measuring jigs for bend, cast etc, gauges for shaping the wood, draw knives for cutting stocks to size and many other tools. They also used the blacking technique to get an exact fit of the action to the stock head. The finisher made all his own 'turn-screws' — screwdrivers to the layman — because the slots of the pins used in gunmaking differ in many cases. Spring clamps were made for compressing leaf-springs into position. All these tools and processes followed in the past apply equally as well today.

This was the basis of the building of a gun, and the organisation of the factory workforce in the time of James the Younger in 1880. As described in Chapter 2, he built Audley House in South Audley Street to accommodate both the shop and a large proportion of his workforce and, apart from this, there was the main factory not far away at 37 North Row, situated in Avery Row just south of Oxford Street in premises rented from the Snelgrove family who later became part of Marshall and Snelgrove, the famous department store. Approximately 120 men were employed between the two buildings and of this number very few were apprentices, the latter only being trained for departments containing older men approaching retirement. Thus only eight boys appear in the records in the late 1880s.

Each man worked at his own bench which he constructed himself to hold his tools and his vice. The benches were lit by gas jets which not only provided light, but heat for tempering metal. There was no electric light either in Audley House or Avery Row until 1900.

The running of both the South Audley Street workforce and the main factory was completely in the control of the factory manager, a Mr Wheatley, known as 'Sheriff' Wheatley by the men for his authoritarian attitudes. He had the power of recruitment and dismissal and was respon-

A group of Purdey workmen in 1889. Standing second from the left is G. Acton (barrel-maker), who signed the letter to James the Younger asking for more pay

sible to the Governor for the organisation of the workforce and the quality of the guns. Under his supervision the 'viewer', Charlie Wilkes, had the difficult task of inspecting and passing the various 'pieces' finished, for Wheatley ran the factory with an iron hand.

However, things did not always go entirely smoothly. After the invention of the Beesley action in 1884, the men were trained to make this and the ejector mechanism invented by Thomas Southgate which accompanied it. Sir Ralph Payne-Gallwey wrote that, by 1887, 65 per cent of all new Purdey actions were being built on this principle and in his book *Letters to Young Shooters* (1890), while discussing the various forms of ejector mechanisms produced by gunmakers in the trade, he praises the Purdey variety for projecting when the gun is opened in an unfired state, a full quarter of an inch: 'Mr. Purdey's ejectors are particularly good in this respect.' A letter sent from Sandringham states that the Prince of Wales too is pleased with the performance of his new ejector guns. However, as a result of these alterations in design, dissatisfaction spread over rates of pay. The rates for the work were in fact high as the Purdeys always employed the best men and wanted to maintain their workforce; nevertheless, in

September 1888 both actioners and barrel-makers wrote to James demanding a revision of their rates and conditions. The unsigned letter from all the actioners reads:

> Dear Sir,
> We, the Piece Actioners, desire to call your attention to the following. When the Action is finished and delivered to be subjected to a good view, the smoking of the sides of the front part of the steel lump to be left for second view. The face of Action not be wasted.
> As we have had no allowance from the extra work for the circle of steel lump, we think 5s on all Actions.
> For the steel fore-ends for ejectors 2s 6d extra and 2s 6d on Solid Bar actions for jointing locks out of sight cracks. Please oblige us with a list of the present price of all Actions.

The barrel-makers' petition, although not so preremptory, shows resentment that other branches of the factory are doing better than they are, and underlines the jealousy felt against the 'day men' as opposed to the permanent staff:

> Sir,
> We, the Barrel-makers in your employ, whose signatures follow below, wishing to approach you on a matter deeply affecting our Welfare, take this means as being the most convenient in the first instance, although we should like you to see us personally upon this subject.
> Firstly we would point out that the money deducted for making tubes is considerably too much, for instance, there is hardly a part that does not require rough-filing and the work generally overhauling.
> Secondly, the ribs have to be re-run after boring for shooting and the new loops having a flat butt piece attached, takes about three hours longer to fit than the old one did, all of which is extra work.
> We should like to call your attention to our position as compared to other branches of your workmen, and also point out that our money compares unfavourably with the Day Men (Barrel-makers).
> In sending you this statement of our wishes we are sure you will believe we are animated by a first desire to improve our unfortunate position and not by a wish to injure anyone. With much respect, we subscribe ourselves.
>
>   Your faithful servants,
>   G. Acton
>   John Hutchinson
>   E. H. Hodges
>   G. Gear

P.S.   We would suggest that ten shillings would be a fair sum to deduct for making Tubes and two shillings a reasonable allowance for resoldering ribs.

Day men were not, in fact, accepted by the craftsmen as being part of Purdey's as the following instance shows. As there was no provision for men losing money due to illness, a Sick Club and Funeral Club was founded in the factory in 1881, James Purdey making an initial contribution to start it off, and each craftsman contributing as much as he could afford per week. A committee was elected from the factory to run the finances; a medical certificate was obligatory before a payment could be made, any member violating the rules of the club could be fined 2s 6d, and expulsion was provided for in the Articles. There were eighty-three members in 1881, and a request was made by the twenty-five or so day members that they be allowed to join. When this motion came up for consideration the vote against was unanimous.

An argument among sportsmen over the merits of fluid-steel barrels as opposed to Damascus barrels was in progress just at the time the barrel-makers presented their request, so it is of interest to read James Purdey the Younger's views on this subject. He was strongly in favour of the former but Sir Joseph Whitworth supplied him exclusively with the steel tubes. Purdey used to say, 'weight for weight steel is stronger than iron and shoots harder, though not of so handsome an appearance as Damascus barrels'. For Damascus he preferred those of Belgian make: 'Not that when thorough sound English Damascus can be obtained they are not superior, but because Belgian workmen are more careful than English, and there is thus less risk of slag and rubbish getting into the Welds.' These tubes were therefore imported from Belgium and made up into barrels in the factory. The only break in this Whitworth tradition was in 1898, when the steel workers went on strike. Between July and December of that year Purdey's had to go to Krupp for barrels and made eighty-three guns with their tubes. James Purdey apologised to his customers and offered to replace all these barrels with Whitworth tubes, free of charge, as soon as the strike was over. Whitworths were eventually taken over by Vickers, and then by the British Steel Corporation.

From the photographs in this book it can be seen that the Purdey workmen of the last decade of the nineteenth century were great characters in their own right. Most men walked to work, and thought nothing of walking five miles both to and from the factory each day. They were very smart on the journey; top hats and bowler hats were worn, with well-cut suits. Once in the factory they changed into their working clothes until it was time to change back before going home in the evening. As the pubs were open all day long there were plenty of stopping places for refreshment.

Purdey workmen dressed in the clothes in which they came to work in the late nineteenth century

Apprentices would be detailed by the men who wanted to eat in the factory to go out to the butcher's, and buy the meat each man wanted. When it arrived they cooked their meals over the gas jets on their benches using 'lard oil' as cooking oil, some men actually using pure olive oil for the same purpose — olive oil, in those days, being used as machine oil. Those men who went to pubs for their lunch, on return to the factory in a happy state would sleep it off on the platform under their benches. As they worked piece work they suffered financially if they did not work for two hours or so, and they made up for the lost time by working late in the evening.

There was a rigid ban on drinking and smoking inside the factory. Drinking, for obvious reasons, and smoking because of the danger of fire due to the presence of small amounts of gunpowder and cartridge materials used for testing guns in 'the hole' on the premises. Snuff was therefore taken in large quantities; and each gunmaker had a small container of snuff on his bench from which he, and a few chosen cronies, helped themselves when they felt the need. During World War I, should one man help himself from another's supply of snuff when passing his bench, the accepted acknowledgement was to become 'bugger the Kaiser' and then a hearty sniff. Drink, however, was occasionally smuggled into the factory in spite of all precautions, and on one occasion Wheatley, returning from a discussion with the Governor at Audley House, found most of the craftsmen either drunk or dead drunk at their benches. 'Where the hell did all this come from?', he shouted, and ordered all the doors to be locked each day from then on, going round the building personally to make sure this order was carried out. The men quickly got round this by lowering buckets on ropes from the windows when Wheatley was otherwise occupied, and signalling to the pot-boy of the pub opposite with whom they had a working relationship. He filled the buckets with bottled beer after taking the money which was in the bottom of the pails. Wheatley in fact was on shaky ground as he owned a pub himself, and was alleged to hand out more work to those who patronised his premises than to the more sober workmen who did not. Eventually this scandal became so great that Athol Purdey had to send for him and deliver an ultimatum that either the pub or his job at Purdey's had to take precedence; the job at Purdey's won the day.

No one was allowed into the factory unless he was an employee and the one private telephone, between the factory manager's office and the lobby in Audley House, was used only by the Purdey family. The factory manager presented himself to the Long Room at 12 noon each day for reports and instructions. The Purdeys in fact kept a close eye on the organisation of the factory and the quality of the guns being built. James, from early in his career, was very careful to see that his customers had stocks made to fit their individual arm-length and stance. His father had used a gun with an alterable stock 'as far back as he could remember', but he felt that the proper fitting of a gun had not had the study it deserved and he made this his aim. The reputation he achieved was what brought in many of the great shots and their friends; though when young Lord Coke tentatively suggested that he thought he required more cast off, he was told that he would get the gun Mr Purdey thought fit to build for him.

The basis of the Purdey gun was that it should be strong, reliable, as light as possible for safety and recoil, the engraving neat and unostentatious, and the whole gun to balance and handle for the individual. These qualities gave Purdey guns their name, but there was something else besides — the confidence the many clients felt in the proprietor and his workmen, and the respect in which he was held by the rest of the gun trade. A newspaper article of 1897 says that during the unfortunate ill-feeling between the Birmingham and London gunmakers, the former had not uttered a word that could in any way reflect on the credit of Audley House and its owner. 'Will you continue to advertise Mr. Purdey by expressing your admiration of the man and your opinion of his work?' was the journalist's question. 'An honest man must, or else say nothing', was the reply from a Birmingham maker of the highest repute. And, of course, James Purdey also had style, and some of this style rubbed off on the guns he built and the men whom he taught to build them.

He charged a high price for his product knowing it to be the best. This price however drew criticism from the influential *Land and Water* sporting paper in an article of 6 July 1889. After remarking that Mr Purdey had inherited an ample fortune from his father and had added to that fortune from the proceeds of his gunmaking it asked: 'Does he not think that it would be a grateful act of gratitude to the Public, his benefactors, to lower somewhat the price of his guns which are at present the most expensive in the Market?' James's comments are not recorded, nor are the comments of the craftsmen whose rates of pay, higher than those in other gun firms, depended on the price at which the guns were sold. Customers did not go to Purdey's in order to economise; they went there and paid the prices asked as they knew that by so doing they personally would benefit from all that the art of gun and rifle making could achieve. Mutual respect between the Governor and the men maintained an extraordinary family feeling in the firm, and the standard of work required. It was therefore of great importance to the Purdey family that a feeling of continuity should exist between the men in the factory and their factory manager, and that they themselves should have full trust in whoever was in control of the workforce.

James and Athol thought they had found just the right man to bring on as the future factory manager — one who would maintain this relationship and keep up the standard of craftsmanship they demanded — when in 1885 they picked Thomas Lawrence, an actioner, who had worked in the factory and machine shop in the basement of South Audley Street for some time. His work was outstanding and his temperament quiet; but his wife died in childbirth and, when the baby died shortly afterwards in 1890, Lawrence, in a mood of total despair, threw up his job, left the country and sailed for India where he became manager of the Army and Navy Stores in Bombay.

In spite of the tragedy of the situation, this infuriated the normally kind Athol. His temper had been getting progressively worse, due either to the combination of Mabel, Julia and the 'Old Man' or the fact that he was suffering from high blood-pressure; so that when in 1898 Thomas Lawrence's younger brother, Ernest Charles Lawrence, applied for a job as a finisher in the factory, Athol, realising who he was during the interview in South Audley Street, ordered him off the premises! However, before Lawrence could reach the front door, Athol called him back with: 'If you are as good as your brother I would be a fool to lose you', and put him in the finishing department. Ernest Lawrence had started his gunmaking career as an apprentice to E. C. Hodges, an actioner to the trade in general, who worked

Ernest Charles Lawrence, factory manager 1914-46

in the Islington area. He had hoped to continue with Hodges but, shortly after he had completed his apprenticeship, the latter gave up making actions and went into business with Henry Atkin. Lawrence, therefore, joined the great gunmakers, James Woodward and Sons, who then had their front shop in Blue Ball Yard off St James's Street. The Woodwards built excellent guns and their name stood very high among the shooting élite, so Lawrence could not have made a better choice. Their rifles were noted for their design and accuracy, as Lawrence used to recall. On one occasion a rifle was returned from the shooting grounds after testing and the tester had inadvertently left a round in the chamber. The assistant in the front shop picked it up and, before checking the chamber, pointed it at a cab horse in the rank in St James's Street exclaiming, 'I could kill that horse', and pulled the trigger. The horse fell dead between the shafts in the middle of the street. The resulting fracas with the cabby became legendary.

Lawrence started with Woodwards as an actioner, and after a few years work in that department was entrusted to make their single trigger. After making a great success of that mechanism he transferred to the finishing shop, where he worked until he approached Athol for a job in Purdey's. He was a brilliant craftsman. His actions were perfect, his knowledge of all types of gunmaking outstanding, and the 'finish' he managed to achieve on a stock was exceptional. As a result of this craftsmanship the men in Purdey's, if their work went wrong, tended more and more to go to Ernest Lawrence to have it put right. Athol spotted this and gave Lawrence the difficult work to complete and, as trust increased, he became assistant to Wheatley. In a factory of craftsmen the worker who can do other men's jobs better than they can themselves is king, and Lawrence soon got that reputation.

A Purdey factory outing at Maidenhead, 1925

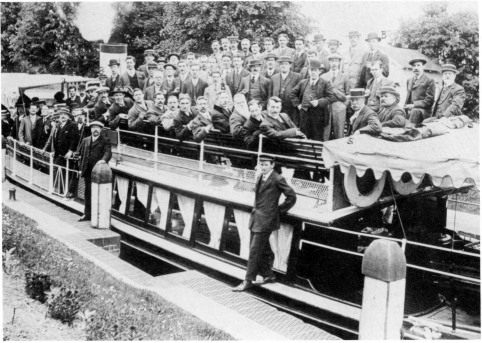

His skill in getting a lovely polish and finish on a stock meant that the other finishers kept having their stocks rejected as not being of the same quality. Wheatley discovered that the reason for Lawrence's success was that the latter made up his own solution from a secret formula of waxes, oils etc, and he thereupon demanded that Lawrence hand over the secret to him. So rude was he in his commands that Lawrence refused point blank to comply and told him that he would supply the factory with his potion in one pint bottles at 7s 6d a time. He called the mixture 'Slacum', and we use the same formula to this day.

The move to Praed Street during 1901, was of benefit to the majority of men who lived in the Edgware Road/Harrow area. As many of the craftsmen as possible were moved in from South Audley Street and Avery Row, and in some cases their benches were re-erected on the small landings of the staircases making it very inconvenient for the workman when anyone was using the stairs. But the site did have its compensations for it was situated opposite the Edgware Road end of the Grand Junction Canal, and conveniently close to the 'Westminster' public house. A large 'well' formed the centrepiece, the barrel-shop and furnace being on the ground floor above which was an upper gallery round which the actioners had their benches. Wire mesh had to be stretched over this well to prevent the barrel-makers being stunned should an actioner drop his tools over the front of his bench, and there were great complaints from the action department when the furnace was lit below and the smoke and fumes poured upwards into their faces.

Continuity of workforce was achieved to a remarkable degree because fathers tended to introduce their sons as apprentices and so family names descended from one generation to the next. Delay, O'Brien, Dean, Hughes and Lawrence are only some of the names involved and when Ernest Lawrence's son, Harry, joined in 1914 as an apprentice, the continuity of craftsman/management was assured. But the calibre of such a body of men, each one of whom was a craftsman in his own right, meant that the factory manager in spite of his authority had to show tact and firmness when dealing with the individuals under his control. And, being individualists, it is only natural that stories about certain of the men have become factory lore.

Harry Acton, a barrelmaker, whose father, G. Acton, had signed the petition to Purdey in 1888, was one of these characters. Outside of Purdey's he dressed very well and gave the impression that he owned the firm, but inside the factory he dressed in dirty old clothes and scandalised his fellow barrel-makers by urinating in his tea-cup rather than go to the lavatory which was at the top of the building. In spite of this peculiar habit, he was the builder of all the rifle barrels at that time and Cecil Purdey tested the results of his labours at the shooting grounds.

William Nobbs, who started as an actioner, invented an ejector mechanism and then went on to invent a single trigger for Purdey which was patented in 1894; but he suffered from bouts of recurring depression. At that time the poverty in some areas of London was appalling, therefore when Nobbs felt a fit of depression coming upon him he would enter a train at Liverpool Street Station, having taken a ticket to Stoke Newington. By the time the train had passed through Bethnal Green and other stations, the poverty of the houses backing onto the line being obvious from his carriage window, he would return to the factory in a cheerful mood, as he had seen that there were many more people worse off than himself.

At a later date, 1931, Walter Warren, an engraver, had one of his pictures, 'A Quiet Smoke', accepted for the Royal Academy. Two barrel-filers about that time had very full lives outside the factory. Sam Simons, who was still working in 1937, was an expert bowls player. He had his cap for Middlesex and on one occasion played for England. Bill Hill, an equally good barrel-man, was Conservative Councillor for the borough of Acton from 1925 until his seventieth birthday in 1931, and was a director of the Hampstead Building Society for eight years.

Even the Purdeys themselves did not escape the effects of their craftsmen's confidence as when one actioner, Tommy Davidson, spoke his mind to Tom Purdey at Christmas 1927. After having a few drinks in the Long Room, Tom and his brother James went over to the factory to deliver seasonal greetings. When Tom arrived at Davidson's bench the Scotsman said: 'Well, Mr Tom, you only come and see us once a year, and when you do you're pissed! It isn't fair!' Tom, completely taken aback, retired speechless from the fray.

Sometimes the boot was on the other foot. On one occasion a senior actioner, Fred Williams, broke the strap of an action on which he was working and was sent across to South Audley Street. Athol Purdey was so angry that he hurled a chair at the terrified man. The noise of it hitting the door could clearly be heard in the front shop, and Williams turned and fled from the building, running all the way back to the factory in Praed Street half-way up the Edgware Road. On arrival in the manager's office he said that he would never again go over and face Mr Purdey 'whatever happened'.

Apart from controlling a workforce of such individual skills and character, the factory manager had to produce as many guns as required by the Governor. Wheatley, with his factory staff of 120 men, produced between 290 and 305 guns and rifles each year from 1890 to his retirement in 1913, delivery time being eighteen months except in the cases of royalty, English or foreign, when the waiting time was cut to nine months. Ernest Lawrence became factory foreman in 1912 and, when Wheatley retired, was appointed factory manager. He hoped to produce a gun each working day

The action shop at Irongate Wharf Road, Paddington, in 1933

and in this he was successful up to the end of July 1914, but on 4 August when the German army invaded Belgium, the situation, as we have seen, changed completely.

But the well-tried system of training has been maintained in war or peace, the only alteration being the reduction of the apprenticeship from seven years to five years as a result of the raising of the school leaving age to sixteen in the late 1960s.

TO THE PRESENT DAY

The position in which Lord Sherwood found himself on 4 March 1946 was similar to that of most chairmen of small companies at the end of the war except that Purdey's was a firm dealing in a 'luxury' trade, building a limited number of highly priced articles each made to the specific requirements of one individual customer and, furthermore, limited in production until apprentices could be retrained to make the actions and mechanisms peculiar to its own designs of gun.

Tom Purdey was over-optimistic regarding the speed at which the firm could make the transition from wartime toolmaking to gun-making when he announced, at the first board meeting after Lord Sherwood's appointment as a director in 1946, that the firm would soon be able to produce 160 guns per year. At that time the gun side had only just restarted work and consisted of the following trained gunmakers: barrel men 4, actioners 4, stockers 4, finishers 5. The total work force including Harry Lawrence was eighteen. The engraving of the guns was carried out by Mr Harry Kell of Poland Street, Soho. He had engraved Purdey guns before the war and, with his staff, carried out most of the engraving for the London gun trade.

The gunmakers very quickly got back into their stride in the various departments for, as Alf Harvey, the 'Royal' barrelmaker, pointed out, they had all been so well grounded in their crafts by their old gaffers during their original seven-year apprenticeships that going from gunmaker to toolmaker at the beginning of the war, and then back to gunmaker again, created no difficulty. The only change was that many of them had joined the Amalgamated Engineering Union during the war while they were employed in engineering and on their return to gunmaking were told by the union that they could not remain members as the AEU knew nothing of the art of gunmaking. They therefore ceased their membership on their return to their previous occupations.

The future for the gun side looked good, as the number of orders that had been placed with the firm during the war amounted to 136, and 158 guns had been ordered between 1945 and 1946 making a total of 294 guns. The price per gun agreed on by Tom and the previous board had, as already stated, been £230; but after Lord Sherwood took over and more realistic higher wages and salaries were paid to the staff and workforce, the costing was adjusted and every customer was written to again and told that the price was now to be £280 for a game gun and £290 for a pigeon gun, 'such prices to be subject to fluctuation up to the date of delivery, in view of the unsettled labour and wage conditions existing at the present time'. This was accepted by most of the customers. However, a far more difficult situation to deal with was the Labour government of Clement Attlee and its Socialist policies. The effort to rebuild a highly skilled small business was not helped by the company taxation imposed on all companies at that time, and the even greater burden placed on the individual wealthy customers who would have to pay for Purdey products. Not only was individual taxation in the higher income brackets 19s 6d in £1, but the Chancellor of the Exchequer, Sir Stafford Cripps, suddenly imposed a 'once and for all' special capital levy on all fortunes over a specified amount; the difference between capital and income being hard to define. Whatever the reasons for this measure, whether social or political, the immediate effect on Purdey's and its craftsmen was disastrous — coming on top of the high taxation which the individual already had to bear, this levy proved to be the last straw. Out of the 294 guns on order, which had ensured, on Tom's estimate, about two years' work for the men just when the firm needed orders to get back on its feet, 116 were cancelled between October 1945 and September 1946 — 39 per cent of all orders. The whole future of the gun firm was in doubt and the livelihood of the very men which the levy was supposed to enhance was immediately at risk. The effect was felt abroad; overseas customers losing confidence in the firm's ability to complete their orders. Major Victor Seely, Lord Sherwood's brother — who with Tom's old friend from the Royal Flying Corps days, Harold Balfour, now Lord Balfour of Inchrye, had recently joined the board — travelled to Lisbon in 1946 in an attempt to persuade Messrs Dias, de Costa, das Corto Silvena and Sousa not to cancel the twenty-three guns they had on order. He was successful and the Portuguese continued their support at this particularly difficult time, even agreeing to pay a deposit of £100 per gun in confirmation of each order so as to help finance the firm over the quoted eighteen months' delivery. The English, on the other hand, who had never before been asked for this form of advance payment, were generally indignant at such a request considering it close to impertinence!

The crisis in industry in general was acute and, with very few orders for the steel-mould side of the business, and the gun side with small prospects of new orders, the skilled men in the mould section were engaged on maintenance, whitewashing and painting for thirty-five hours a week to keep them employed. And at that moment a further disaster hit industry. February 1947 was one of the coldest for years, with unprecedented snow falling in all parts of the country. The government had failed to build up sufficient stocks of coal to meet the needs of industry or of the population in the big cities, and the snow was so heavy that the railways were unable to deliver coal from the pits to the power stations. Total blackouts of electricity followed and on 10 February the power supply to the mould factory was cut off. There being no prospect of a resumption, a meeting of the shop stewards was called and it was agreed that nine machinists would have to be laid off, on the understanding that they could re-apply for their jobs if and when things improved. The rest of the men were put on short time and, as none of the machines could be worked, were kept on maintenance. In the gun side the old Victorian gas mantles were fixed onto the gas jets on each bench, and these gave enough light for the men to continue the hand-craft of the guns.

As a result of these difficulties, two schools of thought developed at board level. Tom Purdey considered there was no future for the gun side of Purdey's. In his opinion the gun was too expensive at £280, the Labour government was determined to destroy all luxury goods and neither the English or the Americans, under the very high taxation they were suffering, would ever shoot again as they used to before the war. Jim Purdey, now living permanently in New Jersey with his wife Mary and his son Bill, wrote a report for the board supporting Tom's views. His American contacts, he wrote, considered that the Purdey gun was far too expensive at US $1,800; the younger generation thought American made guns adequate for the amount of shooting available; and the firm of Abercrombie and Fitch, the acknowledged authority on all sporting activities in the US, agreed. Tom therefore suggested that Purdey's should raise a loan of about £30,000 to buy surplus Ministry of Supply machines and concentrate all the firm's efforts on making steel moulds and precision tools, transferring the gunmakers to the tool side.

Lord Sherwood took a diametrically opposed view. To quote the Minutes: 'In his view he would rather develop the gun side of the firm. He did not feel inclined to go with the "Gauge" business, especially as it would obviously mean the raising of a considerable amount of money, and furthermore the future of such a competitive business was problematical and risky.' Guns were Purdey's business, and just because the firm had done quite well during the war on moulds it did not follow that it could compete in peacetime conditions with much larger tool firms who had known their particular trade for years.

Lord Balfour and Victor Seely suggested a compromise solution whereby an individual interested in property might be induced to invest in the real estate of a factory and lease it back to Purdey's for a joint gun/tool operation. Lord Sherwood did not attend the board meeting at which these points were discussed but the other directors, and Tom in particular, had forgotten the oldest rule of all — 'whoever pays the piper calls the tune'. Lord Sherwood now made his position clear. He said 'that he did not know much about the gauge and tool industry, and probably none of us were very expert in it, and he could not consent to follow a policy which might well lead the company into difficulties in the next few years. He said he considered the company should conserve its resources, and go all out to make a success of the side of the business on which the name of Purdey had been built, i.e. gunmaking'. The arguing was over. Immediate steps were taken to close down the factory in the Great West Road and start to run down the mould side, Harry Lawrence, as works manager, being asked to let the board have a programme as to how he proposed to do this in the most economical way. As the factory was to be closed by 31 August, Tom announced that the bulk of tool orders then on hand could be completed before closure, the remainder could be taken on at the new premises in an old warehouse off Praed Street at 20-22 Irongate Wharf Road. Twenty toolmakers would have to be made redundant, and this was done.

During 1947, only forty-two orders were placed for guns, two of which were ordered by King George VI as a wedding present for Lieutenant Philip Mountbatten. The King told Tom to take his future son-in-law down to the shooting grounds, teach him how to stand properly, and then take his measurements. The King of Afghanistan ordered a pair, as did Prince A. Pahlevi of Persia and, in the following year, during which fifty-three new orders were placed, the Shah of Persia ordered two shotguns and two rifles.

The financial position was now very bad. In November 1947 in fact Lord Sherwood temporarily lost his nerve and asked Tom to prepare a paper showing the chances of the firm making money in three years' time, as it seemed there was no possibility of its doing so in the immediate future. He thought it unlikely that the bank would continue to offer an overdraft, as there was a distinct possibility that the government would restrict banks from loaning money to 'luxury' businesses. So worried was he by the doctrinaire nature of the government's financial policy that he feared only certain classes of goods would be allowed to be made in the future, and only exported to certain countries under proposed currency restrictions. His suggestion that Tom sound out the rest of the gun trade as to their views on future prospects gave the latter the opportunity to point out how wise he had been to insist on keeping a part of the tool side together, and Victor Seely had to step in to keep the peace; as a result the mould side remained in being until 1960. When pressed, Tom admitted that he hoped to

build seventy-two guns each year for the next three years, and if orders for this number did not materialise he would put the men onto building stock guns. In his opinion, and that of Lord Balfour, reconditioned secondhand guns could be sold at £200 per gun, and if the firm could buy, recondition and sell 150 of these weapons in a year, the resulting profit should be in the region of £7,500 per annum. Again financial backing would be needed to purchase the old guns in the first place, and Lord Sherwood eventually made a further loan to the company to set this scheme in motion.

The net loss for the 1947 year's trading was £4,700 — an improvement of only £300 on the preceding year. But Lord Sherwood, now confident again, was certain that his policy of gunmaking was right and instructed that more apprentices be taken on. In November 1946 he had insisted on a scheme being drawn up to recruit more boys, and in December of that year, six boys were accepted as future apprentices, one of these being Lawrence Salter, who had joined as a machinist in the tool side in 1943, and was now apprenticed as a gunmaker, his first real step to becoming managing director in 1971. One more sign of confidence was the already mentioned arrangement to take over the name of James Woodward in September 1948.

It was at this time that Purdey's started to take in for repair and maintenance guns of other makes apart from their own, a policy which up to that time had been deplored by everyone in the firm. Not only did they consider that other guns were of inferior design and workmanship, but it was feared that should another maker's gun go wrong after Purdey's had worked on it there might be difficulties with the original maker. Nevertheless the policy was adopted out of sheer financial necessity.

Moulds were being made for Gillette and Saunders Roe but, as Lord Sherwood had foreseen, competition was acute and price was all-important. Even more serious was the question of wage increases between the two factories. In October 1948, a national award of 5s per week was made to the mould and tool industry, and as a result Tom received a deputation from the gun factory to be treated in a similar manner on the grounds that it was a general cost-of-living increase. Tom explained the whole position of gun orders and costs to the men and said that the gun firm could not be bound by increases given to the mould industry. Being Tom, he went further and said that if the two factories had still been separated by the length of the Great West Road, instead of being side by side in Praed Street, there probably would not have been the same feeling about the matter; in any case, he could not afford this increase. He did not, however, confide the board's fears that unless the orders for the gun side picked up pretty soon some gunmakers would have to be laid off. As everything said in secrecy in the boardroom of a gunmaker's seems to be common knowledge in the factory

Sir Hugh Seely Bart, 1st Baron Sherwood; director 1946, chairman 1955 next morning, Lawrence Salter could tell me years later that he well remembered the feeling in the factory at the time and how the men understood the threat perfectly well!

Tom was now sent off to New York to renew the firm's prewar contacts. He sailed in the *Queen Mary* in February 1949 and spent two months visiting the gun stores in the eastern cities. He took ten orders but wherever he went he was told that his real chance of American custom would be in Texas as although the price of a Purdey game gun — $1,800 after taxes and import duties — was too high for the normal American, the oil-rich Texans would be able to afford it. He reported this to the board on his return and it was agreed that he should go to Texas the following year and see what could be done. A further ten guns followed as a result of his visit, and that gave a total of forty-five orders for the year.

It was at this point in the firm's history that I joined it. Having left Eton at the age of seventeen and a half to join the Royal Naval Volunteer Reserve as an ordinary seaman, in December 1943 I found myself drafted to Butlin's Holiday Camp, Skegness, then known as HMS *King Arthur*. Training followed at Alsager near Crewe, and then the lower deck of the training cruiser, HMS *Dauntless*, from Rosyth. Ghastly train journeys in blacked out corridors, sleeping on kit-bags, being shunted about in sidings, were suddenly relieved by the circumstances of my first leave, when the train in which I was travelling to Newcastle slowly pulled into the platform. As an ordinary seaman, I had been standing all night in the corridor of a first-class carriage, the compartments occupied by senior officers of all services. As the train pulled slowly to a halt they started hailing the one porter, but he had seen me already and was running beside the train, smiling and waving, having been told to take me and my kitbag to my father's car. My kitbag was placed carefully in the boot of the long-nosed Rover, I tipped the porter one shilling, which was my pay for a day, and we drove out of the station past the line of captains, commanders and lieutenant colonels who were still waiting for their bus. That was one of the moments of the service I can say I really enjoyed! After that, there was a course in HMS *King Alfred*, at Brighton, promotion to midshipman, a navigation course at Greenwich, and then long months in 'digs' in Glasgow waiting for my ship, the new destroyer *Comet*, to be completed.

After various commissioning trials we eventually left to join the fleet at Scapa Flow, sailing straight out of the Clyde into a Force 7 gale. The ship heaved over a wave, so did I; and for the rest of my time on board I never recovered from seasickness. We left Scapa Flow to be 'guard ship' in Wilhelmshaven; then on to Japan in January 1946 (touring the ruins of Nagasaki from the top of an open lorry), down to Hainan Island and then through the Timor Sea to Fremantle in Australia. On our return to Hong Kong I was given command of a Harbour Defence motor launch on piracy patrol, with a crew of one leading hand and seven sailors, plus a small Chinese boy of eleven years called 'Cheese-eye' who did all the cleaning, washing etc, and was bullied and befriended by every member of the crew. For six wonderful months, in company of three similar boats, we patrolled between Macao and Hong Kong, and up into the bays round the New Territories until on one awful morning, leaving our berth in the boatpool, I failed to realise the speed at which an oncoming motor-propelled barge was travelling. The resulting collision led to the end of the command and a court of inquiry; the court of inquiry led to a shore job in Yokohama. After eight months in Japan, it was home in a troopship which fortunately stopped in Bombay, giving me the chance to travel by train to Delhi to stay for ten glorious days at Viceroy's House with my elder brother, who had been Air Force ADC to Lord Wavell, and had been asked to remain with Lord Mountbatten during his viceroyalty. Back to England in a York aircraft, being airsick at every bump from Basra to Cairo, then to a Fleet Air Arm station at Everton, near Inverness, until demobilisation eventually came in early 1947.

After demobilisation I had no idea of what I wanted to do. Then suddenly I was offered a job by Sir Walter Monckton. After the partition of India on 15 August 1947, Hyderabad and Kashmir were of such size that they were in a position to negotiate with the new governments of India and Pakistan as to their futures. Walter Monckton was acting for the Nizam of Hyderabad over these negotiations and his envoy was John Peyton, now Lord John Peyton. The aeroplane carrying John Peyton to India had recently crashed and he had suffered severe concussion, so it was suggested that I took his place as Walter's personal assistant. As a result we travelled to India and Pakistan on four different occasions during 1948, staying in Hyderabad and Viceroy's House in Delhi, until in July I travelled to Hyderabad alone in order to transcribe and bring back a letter the Nizam had written to King George VI asking for help in presenting Hyderabad's case to the Security Council in an attempt to prevent an attack by India on his state. On my way back through Delhi, I was detained and the letter was opened and read. All hell broke loose, and the Indian government's 'expressions of regret' over the incident did not carry much weight, especially as on 26 July Pandit Nehru announced to a mass-meeting in Madras that India's aim over Hyderabad was 'full accession or else the disappearance of the State'. On 28 July the Nizam sent a telegram to Delhi demanding the withdrawal of Indian troops from his borders, and on 18 September the Hyderabad troops eventually laid down their arms after four days of heavy fighting. The appeal to the Security Council was dropped, my job with Walter Monckton naturally came to an end, and other employment had to be found.

Regretfully after all this foreign adventure, the 'other employment' was that of a shipping clerk in Billingsgate. It seemed very mundane. However, one good thing was that my uncle, Lord Sherwood, knowing that I liked shooting and in gratitude to my father who had helped considerably with several of his family's affairs, had promised to give me a small duck pond surrounded by a rough field of about two acres, and containing an old keeper's hut. In early 1948 it suddenly came to light that this small property had been sold with other land. My father was so incensed that he determined that his brother-in-law should replace this gift with another; so he travelled down to Brighton and bearded Lord Sherwood in his flat in Chichester Terrace. The first thing I knew about these arrangements was early in 1949 when the telephone rang and I was told, 'Uncle Hughie has given you Purdey's.'

On joining the firm, I was sent to the factory in Irongate Wharf to watch the various processes by which the guns were built, and receive a basic training from Harry Lawrence. Tom went off to Texas in 1950 to see what he could glean in orders from Fort Worth, Dallas and Natchez. He found that the Texans wanted over-and-under guns built the way they liked, not in the way he advised them; but as there was no top-quality American gun on the market he expected to sweep the field once the oil men had been taught what a Purdey gun really was. Tom took orders for four over-and-unders, two of them 28-bores, and then went on to New York where he made arrangements with Abercrombie and Fitch to handle Purdey guns and sell stock guns at a discount. There was very nearly a major break when Tom flatly refused to grant that firm the sole Purdey agency for America in exchange for more orders, and as a result we have never, except for one very limited period in 1962, granted a sole agency anywhere in the world, and always reserve the right to deal with any customer for whom we build a gun. The trip, producing nineteen orders, was a huge success; Tom, by his character and personality, rebuilding the name of Purdey's after so many years of war. Lord Sherwood was delighted with Tom's stand over the sole-agency controversy and reaffirmed that we were an entirely independent company and must always remain as such. The arrangements made with Abercrombie and Fitch continued harmoniously for many years until the sole-agency issue was raised by them again in 1961. They wrote that they had taken a long, hard look at James Purdey's, and did not like what they saw. Lord Sherwood replied that he had taken a long, hard look at Abercrombie and Fitch and he did not like what he saw; and unless the matter was dropped we would not deal with them again. The issue was dropped.

Tom's trip had changed the whole future of the gun side of the firm and orders now started to flow into South Audley Street once more. He had written to the Duke of Windsor about his old nursery-maid, Charlotte, who had served with the Royal children, and received a reply which confirmed his Texan opinions:

85 Rue de la Faisanderie,
Paris 16,
France.

June 13, 1950.

<u>Private and Personal</u>

Dear Tom,

   Many thanks for your letter of June 5th. It was only proper that the guns your firm built for my father should be mentioned in connection with my statement that "<u>everything about him was always of the best.</u>"

   It has taken three, hard, exacting years to write my memoirs, and now I am trying to meet my publisher's deadline to my satisfaction with a book. But it has been interesting to recall the people and experiences of one's youth, unfortunately relegated to a distant past.

   My mother had Charlotte Bill - or "Lala", as we used to call her - come to see me at Marlborough House in 1945. She is still living for all I know, as is Frederick Finch, our valet and ultimately my butler at York House. Those old-timers die hard.

   I am glad to hear your business survives the Socialist regime, which seems bent upon destroying the fine old "custom" trades in favour of mass production. There is an awful lot of money in Texas, where I hope you were able to sell some of your fine guns. The Duchess and I were fascinated by the Texans, in whom the spirit of adventure is still alive and thriving.

   You must have enjoyed your brother Jim's visit with you. I had forgotten he is now living in America.

   Thanking you again, and with kind regards,

        Sincerely yours,

        Edward

T. D. S. Purdey Esq,
57-58, South Audley St,
London, W.1.
England.

One of Tom's later visits to America, apart from being a great success, was very nearly his last. He was suffering from diabetes and travelled with the necessary tablets and injection equipment. On the evening of 11 January 1952, with Jim and Mary Purdey, he boarded the crack streamliner train *The City of San Francisco* in St Louis to tour the west coast. In the high sierra of Nevada the worse snowfalls for years had fallen, and on 13 January the great train stuck fast in a snowdrift with 226 passengers on board. At first it was all rather fun and everyone expected that the train would be on its way in a few hours, but the blizzard increased and for three days the train remained motionless and snowbound; the water supply and lavatories froze solid and snow blocked the engine outlets so that poisonous fumes came back into the coaches and twenty-seven people collapsed. Heating and lighting failed, and the passengers tied blankets and sheets round their legs to keep themselves warm. The snow was so bad that the army and airforce were at first unable to get through to the rescue, and food started to run out. A coastguard helicopter dropped a supply of eggs all of which broke, some on the head of an army colonel, and Tom ran out of insulin. He became seriously ill and it was only thanks to the attention of a doctor on the train, Walter H. L. Roehil, and Jim and Mary's support, that he managed to survive until they were rescued.

The costing of the gun had to be gone into once more as there was constant pressure for higher pay from the factory. Tom had once again reduced his estimate for the number of guns we could produce per year, and settled on a figure of fifty of all kinds. The costing in March 1950 was as follows:

Average cost per gun		*50 guns per annum*
		£
	Materials	37
	Production Wages	98
	Indirect Expenses (£185)	182
		317
Average selling price per gun		360
Profit per gun		43

After a long discussion with the accountants it was decided to fix the home price of the gun at £300, and the overseas price at £360. The home price was kept low as the English considered the price to be far too high and suspected the firm of making disgracefully high profits; further there was a Purchase Tax on guns of over 30 per cent. So serious was the effect of Purchase Tax on home orders that Lord Sherwood attempted lobbying in the House of Commons to try and get it reduced and Tom was instructed to approach the Gunmakers' Association and attempt a trade appeal to the government to help skilled crafts.

On 29 October 1950, Major Arthur Collins was made a trustee for my shares in the business and a director, and I was confirmed in my commencing salary of £6 10s per week.

In the mould side the profit ratio stood at only 15 per cent as a result of the intense competition from others in the trade, and it was announced that whilst it was realised that the cost of living was rising, bearing in mind that the staff pension scheme had been started in the year 1949 the board had decided in the interests of all concerned to defer for the present any increases in wages. And it was in 1950 that all the effort put into the business since 1946 succeeded in making a small profit for the first time.

New markets were still needed for the gun side and in 1951 Tom suggested that he should travel to Australia in August in the ss *Oronsay*, and return in September in ss *Orcades*. However, he got cold feet and on 14 September I was sent out in his place, travelling in a Constellation aircraft and clutching £250 in expenses. On arrival in Melbourne on 18 September, I was told that no one in Australia would dream of paying the 'huge' price of £360 for a Purdey gun, especially as the sales tax on a luxury weapon was 8½ per cent. On 27 September the Australian budget was announced and the sales tax was increased to 33⅓ per cent. Every gun shop and individual admired the over-and-under and side-by-side sample guns I had with me, but each in turn said that due to the increase in sales tax the purchase of the guns was out of the question.

I flew to Deniliquin, Adelaide, Canberra and Sydney; everywhere it was the same story. But Australian generosity came through in Adelaide when, after telling my woes at a dinner party, Tommy Barr-Smith showed me his old gun and said that if any one of the sample guns had similar measurements he would buy it on the spot in order to help out. One gun was of almost identical specifications so he wrote a cheque at the dinner table and I sold one gun during the month and a half I was in the continent! Tom was off to America again in December travelling in the ss *Caronia*. The sea was so rough that the liner docked thirty-six hours late in New York and the *Queen Elizabeth*, on her homeward voyage, arrived at Southampton with forty injured passengers. Tom never felt seasick and thoroughly enjoyed the crossing, walking the decks and puffing on his pipe to the discomfiture of the other passengers, many of whom must have been feeling dreadful. Harry Lawrence who had been made a director in July 1951, took charge of the factory while he was away.

It was during this trip that Tom built up such lasting links with the American film industry. Douglas Fairbanks was lent a gun in 1952 and wrote to Tom as follows:

FREMANTLE 2685

28, THE BOLTONS,
KENSINGTON, S.W.10.

18th August, 1952.

Tom Purdey, Esq.,
Messrs. James Purdey & Sons,
57, South Audley Street,
London, W.1.

Dear Tom Purdey,

    I have, as you may have learned, returned your gun, gun-case and unused cartridges and I cannot tell you how appreciative I am for your courtesy.

    It is no wonder that Purdey guns are so famous throughout the world and on this, my first experience with a shot gun, I surprised my host and myself by actually hitting a few of the birds I was aiming at; the gun pratically aimed and shot itself with very little assistance from me! I have rarely had a better time in my life and this brief holiday owes its success, in no small measure, to your much appreciated courtesy, advice and help.

                Yours sincerely,

[signature]

DOUGLAS FAIRBANKS, K.B.E., D.S.C.

He also suggested that Tom should contact his friends in the industry on his next trip to Los Angeles, among them Fred Astaire, Gary Cooper, Darryl Zanuch. Tom already knew Charles P. Skouras of Twentieth Century Fox, this friendship with Mr Skouras having started in difficult circumstances when, summoned to the mogul's office in 1951 to show his wares, Tom felt he was being patronised. So when Skouras slapped Tom on the back after ordering a 12-bore and a 20-bore gun, Tom returned the slap with interest almost knocking the film producer off his feet. Skouras immediately asked him to dine and, as at the time he was financing the building of St Sophia's Cathedral in Los Angeles, took Tom along after the meal to see the building and switched on the carillon at 11.30pm, waking half the city.

Meanwhile, at home, the firm was still loading cartridges by hand in the shop designed for the purpose in 1880. In 1949, 1950 and 1951 some 620,000, 787,000 and 658,000 respectively were loaded in this way by Ralph Lawrence and his staff. Due to the continued shortage of materials of all kinds, a form of rationing was still in force for all customers and this was to continue for a further five years. However, new powders were now being introduced by the metals and explosives branch of ICI and their latest, '56', was in production. This powder, which was to take the place of all the other 'individual' powders such as EC, Diamond Smokeless and Schultze, was considered too difficult to handle in a residential area, and so in 1951 hand-loading was stopped and Imperial Metal Industries (IMI), a subsidiary of ICI, took over the loading of all our cartridges. The 'crimp' closure of cartridges was now introduced in place of the top-wad. This innovation was treated with great suspicion by our customers and in 1952 a letter had to be sent with each new consignment explaining the alteration:

> We would mention that the cartridges we will send you have the *New Crimped Turnover*, which is the very latest development with Sporting Shot Gun Cartridges. We have carried out most extensive trials with this type of turnover and the ballistics obtained give us every reason to recommend such cartridges to our customers. We would also mention that cartridges with the new Crimped Turnover are cheaper than the old type and that there is little doubt that in 1953 the turnover will entirely replace the old type.

Among other worries of this year Mr Child of Cambridge was handed the gun belonging to Mr Child of Slough, and only discovered the mistake on his return home. And while we received twenty-one testimonials for good work in that year, this was counterbalanced by thirty-six complaints. Three concerned new guns, others concerned cartridge performances, accounts department's mistakes, late deliveries, standards of repairs and rudeness in the shop. On 1 June 1952, Lord Balfour of Inchrye and Tom's lifelong friend, retired from the board due to the other pressures on his time. His work for Purdey's had been outstanding and everyone greatly regretted his leaving.

Early in 1952 a terrible accident occurred. Ralph Lawrence had been 'loaned' to the Gunmakers' Company in October 1951 after the handloaded cartridge side of Purdey's had closed down. His position was that of assistant proof master, and the understanding was that he would eventually take over as proof master at the Proof House, and be responsible for the safety of every gun proofed. On the morning of 18 March he was dealing with a batch of cheaply built bolt-action rifles which had been submitted for proof without their stocks. Having loaded the charge into one of the rifles, he was about to place it in the proof chamber when the telephone rang. He put the rifle down on its 'strap' and must have caught the trigger on his shoe, for as he bent over to pick up the receiver the muzzle of the rifle fell towards him and the charge exploded in his neck, killing him instantly. In November of the same year his father, E. C. Lawrence, the factory manager who had served three members of the Purdey family and the firm so well, and had been born only seven years after the founder of the firm's death in 1863, died at the age of eighty-two.

In order to boost sales, a policy of extending our wares through accessories such as mackintosh coats, boots etc was introduced, and a designer was hired to 'transform' the front shop. This was not a success. The frontshop staff were trained to deal with guns and so had no time for the niceties of the fit of a mackintosh; the older customers protested vigorously that the firm had lost its atmosphere and objected to the 'pansy' new clothing. Worst of all, the prototype of the new raincoat which had been designed especially for Purdey's proved a disaster when heavy rain at the Three Day Event at Badminton ensured that the wearer was the only member of his house party who had to change right down to his underwear before going into lunch.

By 1953 the price of the new gun had increased to £400 with an eighteen-month delivery, and new barrels for old guns (known as 'fitter-in barrels') had increased from £95 to £120 per pair. Customers were insisting that we introduced some form of credit service, and this was accordingly arranged. A hire purchase agreement was also considered and adopted. The scheme was called the 'Purdey Credit Service', and the arrangements were made through the United Dominions Trust; it was considered very go-ahead and modern.

Tom had now been ill for some time and it was becoming obvious that he would not be able to return to the board as a working director. On 1 May 1955 he was made president of the company, granted a pension for life and the company gave him Fern Cottage at Chilham in Kent — a most attractive house next door to the excellent Woolpack Inn and within easy walking of the River Stour, where Tom could indulge his passion for trout fishing. His manservant William lived in the cottage with him and Tom continued to be called on by the firm from time to time for consultation.

As a result of this decision, another representative had to undertake the yearly selling trips to America, and Harry Lawrence flew to North America at the end of October 1955, full of trepidation and doubts as to his ability to sell Purdey guns. He was an instant success, the Americans admiring his great knowledge of gunmaking and being captivated by his courtesy and skill. He took with him, besides the measuring gun, the miniature gun he had made in 1935; this additional proof of his skills took the Americans by storm and he took thirty orders in four weeks, and in the years that followed built on this first great success.

On his return a serious difficulty had arisen over the training of apprentices. National Service was in full swing, all able-bodied boys being required to serve for two years with one of the armed services from the age of eighteen. This was four years after they had started their apprenticeship at the factory and just when they were beginning to find their feet in their particular branch of gunmaking. Even if they did return after their period of service, it was very difficult for them to settle down once more and take up their craft where they had left off. They would have to relearn their basic skills and, at the age of twenty, would be working alongside boys of seventeen. Moreover, they would not be coming out of their 'time' until they were twenty-three, whereas men not fit enough to join up were finishing their apprenticeships at the age of twenty-one and then going on to a higher basic rate. Harry Lawrence spent many anxious, depressing hours trying to get exemption for those boys he knew would be good craftsmen in the future if only they could be given the time and the peace to continue their training. After all the effort he had put into rebuilding the firm from the mere seventeen men available at the end of the war, it now looked as if his training programme was in jeopardy.

A brighter note was that Her Majesty The Queen granted us her Warrant of Appointment for guns and cartridges in 1955, and the Duke of Edinburgh his for guns in 1956.

It was now that Tom Purdey fell very ill. Cancer was diagnosed and he went into hospital for an operation. There was no hope of recovery and in March 1957 he died. After a service at Chilham Church he was cremated at Ashford Crematorium and his ashes were placed with those of his mother and father, Mabel and Athol Purdey, in the crematorium grounds, rose trees being planted on the spot. On 29 March a memorial service was held in the Grosvenor Chapel, where for so many years Tom had served on the church committee. And so ended the Purdey family's managing link with the firm. Jim was still a shareholder, and his interest would pass to his son

Jok; but neither of these two members of the firm founded by their ancestor in the year before the battle of Waterloo would take part in the future running of the firm although they both remained very close to its activities.

Poor Tom had done very well during such a difficult time. While still a schoolboy at Eton, he had been left two-thirds of the business over the head of his elder brother, a situation which could have been disastrous had it not been for the great affection the two boys had for each other, and the tact displayed by Jim when he joined the firm as junior partner to his younger brother. There was nothing Tom could do to bring the fortunes of the firm back to the level it had enjoyed before World War I. Falling markets, civil unrest and social change, his mother's appalling extravagance, the Depression followed almost immediately by World War II, thwarted his schemes at every point. The amazing thing is that the firm was still in being and starting to expand again at the time of his death. He maintained the name and standards of Purdey even though four other families controlled the shares during his lifetime, each one with their own ideas of how things should be done, and it is mainly due to him that the firm still keeps its family links and personality today.

He was a very loyal and affectionate friend, an artist and musician, a passionately keen fisherman, an average shot, but above all he understood gunmaking, and his contribution to safety and to the standards still held by English gunmakers is immeasurable. He was fiercely independent, refusing to build guns for customers he did not like, and any criticism of the guns he made, especially from a member of the gun trade, would start a magnificent tirade about the antecedents and habits of the man concerned. His eccentricities and humour made working in Purdey's the greatest fun, and even the excess of whisky which eventually led to his final illness was in the highest traditions of his grandfather, James the Younger, who had much the same temperament and charm.

In February of 1958, a further grant of wages was authorised for all engineers; once again trouble brewed about a similar rise for the gunmakers. The price of the new gun had, therefore, to be raised to £500. Harry Lawrence voiced the opinion that his chances of taking more orders in the United States, Portugal and Italy would be greatly diminished as a result — the customers there were already complaining at the price — and that in England it would be disastrous, with accusations of making excess profits and encouraging inflation.

The effect of raising the price of the gun fortunately did not have the disastrous results expected as the next trip to America produced an abundance of orders and the delivery period for guns started to be lengthened as a result. Furthermore, in November 1958 Harry Lawrence went to Dallas, Texas, for a British Fortnight held in the Neiman Marcus store and took fifteen orders from new customers, about one order each day. The only cloud on this excellent horizon was the continued desire of the Texan customers for a 'pretty gun', or 'purdy gun' as they pronounced it, which went against the age-old tradition of 'Purdey' quiet engraving. Engravings of quail, mourning dove, ducks and dogs in gold were all the rage but, as James the Younger had said: 'If that is what the Americans want, I suppose we had better give it to them.'

Fortunately we were well placed to meet these demands. On 4 December 1950 we had taken on an apprentice, Ken Hunt, who had trained under Harry Kell; and on 12 January 1952 he was signed up at Purdey's and worked in the Praed Street factory. There had been small demand in the English market for decorated guns, and in order to keep Ken busy and so that he could practise his skills, we started a line of silver cigarette boxes and writing pads on which customers could have their house, dogs or favourite game birds inscribed. Not only was this very lucrative, the boxes selling for £18, but it increased his skills so that when the first 'gold guns' were ready for engraving he took on this work at once. So popular were the first results of his work that new gun orders stipulating special engraving started to flood in and we thought that we had at last found enough work to keep us safe for many years. However, there was a snag. When you can only build seventy or eighty guns a year, and a third of the customers stipulate Ken Hunt's engraving which took up to a month for each gun, not only do you cut your production of guns due to the bottle-neck caused in the engraving department but the finishers start to run short of work as a result. So instead of elation there started to be gloom all round the board room.

This depression was aggravated by the news from the mould factory. For the year ending December 1958, due to the latest wage increases, the net loss was £1,900; and for the year 1959 there was an estimated loss of £4,000. It was now obvious that, far from being the side of the business which should have been supporting and subsidising the guns, this factory had become a liability which could bankrupt the whole concern. When the net losses on moulds reached £7,900 as against the net profit on guns of £10,000, it was obviously time to call a halt and end it. It was therefore decided to close down that side of the business, and notice was given to this effect. Every man in the tool side immediately found another job, a large number being taken on by Gillette, for whom we had made many tools; the final orders were completed and the factory closed in the spring of 1961.

The firm had at last returned to the building of the one thing it knew about — sporting guns and rifles. The Long Room was reorganised, the family portraits cleaned, named and re-hung. As there was no portrait of Tom, an excellent portrait was executed by Robert Swan from photographs, so that we now had a portrait of every Purdey since 1814. A trunkful of old photographs of customers, nearly all signed and presented when they took delivery of new guns, was found in the attics. These were reframed and hung on the walls, and the shields of members of the firm who had been Masters of the Gunmakers' Company were placed above them. The very ugly lights which had hung since 1930 were removed and a new chandelier, made from some of the old Victorian gas brackets which had originally lit the front shop, and surmounted by stags, was made up in the factory and suspended from the centre beam.

Harry Lawrence went off to Los Angeles in 1961 to try to get more orders; a gun was exhibited in the Copenhagen British Design Exhibition, but it was decided not to go to Moscow for the British Trade Fair that year. Before these visits took place the price of guns was raised once more. The over-and-under gun was to be £700 for export and £831 5s for home buyers; double rifles were to cost £650, and side-by-side guns were priced at £550 for export and £653 2s 6d at home. The firm was going to try and live solely on its gunmaking abilities.

In January 1962, I was told by the Spanish Ambassador, the Marquis of Santa Cruz, that General Franco, the Spanish head of state, had been badly hurt in an accident when shooting outside Madrid and was using his pair of Purdey guns. These guns, originally built in 1923, had been acquired by the general in 1959 and were perfectly sound. The left barrel of the No 1 gun had burst open at the end of the chamber so that the force of the burst travelled downwards smashing the wood of the fore-end, which splintered into Franco's hand smashing his left thumb. He was rushed to hospital where his wounds were dressed and he was treated for bad shock. I went to Spain to inspect the gun, and although I was not permitted to bring the weapon back to England I was given every help, and photographs of the damaged barrels were taken which I brought home. From all the evidence it was clear that an obstruction, probably a 20-bore cartridge, had accidentally been put into the gun while loading and this had lodged at the end of the chamber. In the hurry, the loader must have thought he had dropped the cartridge and so had loaded a 12-bore behind it, with the result that high pressure had caused the classic 'ring-bulge' to form before the barrels burst, and parts of the brass and paper case from the smaller cartridge were found embedded in the torn part of the barrel. Fortunately the strength of the action and barrels had prevented any more damage to the shooter, and General Franco acknowledged the fact by ordering another pair of guns as replacement.

After this excitement Harry and I left for America and Japan, forty guns being ordered in the United States of which twenty-nine were to go to new customers. Japan was more complicated as it was necessary to appoint a sole agent who would deal with all questions of finance and import controls. This went against all our decisions of the past but there was no way out, and a Tokyo agency became our agent for a test period of two and a half years after which the agency was not extended. We took eight orders and as a result of these fifty guns, all of which would have to be finished and delivered in eighteen months time, the delivery of all new guns was extended to two years.

Up to that date, all our tubes for barrel-making had been supplied by Vickers-Armstrong. They had a few elderly men working there, straightening the tubes by eye and hand, and making a wonderful job of it. We received a letter in July 1962 informing us that they were closing down the plant, but would continue to fulfil orders for three months. As we always received barrel-tubes on order it had not been considered necessary to buy in large stocks of the various bore sizes; therefore this sudden closure meant that we would have to find a new source of supply or close down, as it is a well-known fact that you cannot make guns without barrels. Fortunately we discovered a letter in correspondence of 1960 in which Vickers had stated that, should they be forced to close through lack of orders, they would give us at least six months' notice of their intention. Confronted with this letter, they very honourably stood by their commitment and kept the department open until they had completed 350 pairs of assorted gun and rifle tubes which fulfilled our requirements for six years, during which time we could find another source of supply. As the British Steel Industry was unable to supply us with the tubes we required, we eventually found an alternative supply through a most reliable firm in Belgium.

On 25 July 1963, Jim Purdey died at Holmchel, New Jersey. He had been ill for some months, and a year earlier had been operated on for the removal of both legs due to thrombosis. He was very brave and used to wheel himself about in a chair round his home and garden. He was seventy-two.

Harry Lawrence was now sixty-four years of age and so was on the verge of retirement. This was, of course, unthinkable, both to himself and the firm. He was as active and knowledgeable as ever, and so a proposal was made by Lord Sherwood at the board meeting of January 1964 that 'The Employment of Mr. C. H. Lawrence (now aged 64) as Managing Director be continued to 1st February 1975, but that he should have the right at any time during that period to give the Company six months notice in writing of his intention to retire.' Needless to say, Harry has never done so and is still with us today in 1984, as is Eugene Warner who is Harry's elder by three years.

By now the effects of postwar taxation and death duties had broken up existing estates, and syndicates of the former farmer/tenants, supported in many cases by businessmen or business firms, started to take their place. These men could afford the best, and so not only were large numbers of pheasants raised for shooting but more guns were needed. It was here that

Harry Lawrence made his greatest contribution and name, in building guns for customers who longed to shoot well but in many cases had little experience of the sport, and his advice and kindness to them all was greatly appreciated. This was especially the case with visiting Americans who naturally knew nothing of the conditions which exist on grouse moors in Scotland or the north of England, and had to rely on his advice for the best gun for their needs. His tact and great gunmaking skill ensured that they always received his best care and attention at all times, and the result is shown in the admiration and trust he still commands in the gunmaking and shooting world.

One of the most difficult things is to explain the differences between shooting in America and shooting in England. The ruffed grouse of New Jersey is 'hunted' in scrub woods; the red grouse of England is a totally different bird. Shooting with two guns in America entails taking out a side-by-side gun and an over-and-under gun with the prospect of firing off a few shots with each, whereas using these two different guns in Scotland would create great confusion to a loader in a grouse-butt. Another even more basic difficulty is the language difference between the two countries — 'gauge' instead of 'bore'; 'shells' instead of 'cartridges'; 'hunting' instead of 'shooting' — and in the last analysis a difference in basic slang, an example of which is cited by Captain Ralph Cobbold. During lunch on a Scottish grouse moor, a keen American guest was discussing the maintenance of his guns: 'What do you do with your guns in the off-season Captain Cobbold?' Ralph answered facetiously, 'Oh, I usually put them in hock.' 'Oh, but Captain', replied the American, 'Doesn't that corrode the barrels?'

Orders kept coming in to the front shop in a steady stream and the factory produced an average of seventy guns per year between 1964 and 1969. The American trip by Harry Lawrence and Lawrence Salter in 1968, designed as much to introduce Salter to the American customers as to take guns, accounted for thirty guns and rifles. But inflation was beginning to tell. The accountants had been urging us to increase the price of the gun for several years, but the fear of meeting sales resistance prevented our agreeing to their recommendations and we decided to make a yearly increase of small amounts of £50 to £75 only, hoping that things might stabilise. This policy proved to be worse than useless as inflation gathered speed, and the call for higher wages in order to keep step increased. American customers saw what was happening and took full advantage of the situation by ordering as many guns as possible at these low prices; but the English complained bitterly at the ever-increasing price and, as we lived in England and were constantly hearing these complaints, we naturally tried to keep in with our fellow countrymen by keeping the price down. The 'swinging sixties' were in full spate and the young men in the factory, seeing that their contemporaries outside were demanding, and getting, higher wages for far less skill, started to leave as they saw no hope of higher wages. As a result of trained men leaving, the rest of the factory became dissatisfied and apprehensive of the future, and so a vicious circle was set in motion. New apprentices were taken on but it would take five years to train them and all the time the pressures on margins were growing.

In the country as a whole it was a time of great change and unrest. 'You've never had it so good', was perfectly true, but the younger men wanted more, and if the firm which employed them would not provide this extra they left. In the case of the trained young craftsmen this was fatal to a firm like ours. Unflattering comparisons of wage rates in the public sector vis-à-vis rates in the factory were always being levelled at the factory manager — our customers were the richest men in their own countries and so why didn't the board raise the price of the gun and make the customer pay more? Unfortunately the customers did not agree with this simple reasoning, and so the turnover of the young men continued.

The increased crime rate including armed robberies was now so serious that the public at large began to protest at the ease with which any individual could buy a shotgun. The sale of rifles and pistols had for many years been subject to the possession of a Fire Arms Certificate authorised by the police; but now the home secretary suggested that some form of licensing should be introduced for shotguns, and a committee of the Gun Trade Association met representatives of the Home Office to work out recommendations for legislation on the issue. Captain Marcus Kimball MP (now Sir Marcus) was chairman, and as a result of Harry Lawrence's suggestion that the meetings should be held in the Long Room as a place convenient for all parties, the committee came to be called the 'Long Room Committee'. After many meetings, and much discussion, a form of control under the Shot Gun Certificate was worked out. This stipulated that before any smooth-bore weapon can be purchased, the customer's certificate must be shown to the gunmaker and the details of the certificate entered in a register subject to spot checks by the local constabulary. Should a gun be brought in for repair, the gunmaker is forbidden to hand the weapon back to the customer unless his certificate is in order. To obtain a certificate, application must be made to the customer's local police station where security arrangements for the weapon or weapons are investigated before a certificate is issued. Many individuals have collections of firearms built up over the years of which they are extremely proud; however should one of these bona fide collections be in the possession of the occupant of a flat in a tower block in the middle of a big town which has a high rate of crime, naturally the local police authority will have grave reservations about granting a licence unless the security arrangements are absolutely to their satisfaction.

The price of the gun had been raised to £1,100 in August 1968 and considered ludicrously expensive by the English and dirt cheap by the Americans who were enjoying the effects of inflation and ordering as fast as they could. In January Harry Lawrence told the board that, as he had 250 guns in hand with a delivery of eighteen months to two years, and as we could only build 75 guns each year, it would be wiser not to go to America on a business trip for at least twelve months but to try to get through the backlog of orders. The delivery date would have to be extended for new orders to 2 to 2½ years and another increase in price must be considered. Even then it was not fully appreciated that as all these orders had been taken on a fixed-price basis, the effect of inflation would be devastatingly dramatic. Advice had been sought from reputable banking acquaintances as to the height inflation could reach and in those days these level-headed and experienced businessmen gave the figure at approximately 7½ per cent or 10 per cent 'at the outside'. As a result, the price of the gun was raised to £1,300, and we hoped that it would not put off prospective buyers.

In February 1969, Lawrence Salter became joint managing director with his uncle. In March 1970, however, the latter resigned this position at the age of 70, though he remained as director and adviser. The Honourable Anthony Tryon was elected a director. By now the rate of inflation had risen disastrously and the effect of taking on so many orders was beginning to show. We had forty over-and-under guns on order with a delivery of two years at fixed-price contracts; but with only two men working on the over-and-under actions, each of which took six to eight weeks to complete we could only build ten of the guns each year. It was therefore impossible to complete the guns in the time quoted or at the price agreed. We decided to write to all the customers who had guns on order telling them the position, and letting them know that if they cancelled we would repay their deposits with interest. At the same time the price was again raised to £1,500 for the side-by-sides and £2,000 for the over-and-unders. Just to help matters along, the landlords of the factory announced that they intended to redevelop the land on which the building stood and make way for the Metropolitan Hotel which now stands on the site in Praed Street. Alternative accommodation was offered at Bishops Bridge Road, Paddington. This meant that, in the middle of this financial crisis, the whole production capacity of the firm would have to be disrupted while the benches, machines and men were moved to the new premises.

And then on 1 April 1970 Lord Sherwood, who had been ill for some months, died in his flat in Brighton. He was buried in the family plot in the graveyard of St Mary's Church at Brooke, on the Isle of Wight. His funeral was attended by a host of local inhabitants and friends although he would have been most annoyed to have known that his wishes for the funeral service had not been carried out to his exact requirements. The hymn 'All

Things Bright and Beautiful' was his favourite; but unfortunately the parson of the parish was away and his locum, a keen young clergyman from Portsmouth, came over. He refused to allow the verse:

> The rich man in his castle
> The poor man at his gate,
> God made them high or lowly,
> And ordered their estate.

to be included. Uncle Hughie's wonderful butler/valet, Alan, who had served him so faithfully for many years remonstrated with the divine: 'Oh, but sir, His Lordship will be most distressed if that verse is left out; he always said it was the only verse in all the Hymns Ancient and Modern which made any practical sense!'

Lord Sherwood was a man of very keen intellect though his eccentricities were famous, such as wearing a coloured skull cap, a multicoloured shawl, and bedroom slippers embroidered in gold with the letter 'S' surmounted by a coronet when he attended board meetings, and even in the House of Lords. His practical common sense and enthusiasm not only saved the firm of Purdey from oblivion in 1946, but rebuilt it over thirty difficult years so that it is still in existence today, an anachronism in a world of ugly machinery, fast efficiency and rather dreary industry. And it was mainly due to his tact that the Purdey family quite happily remained as titular heads of the organisation which he controlled. He was a most entertaining and amusing companion who could put forward extraordinary ideas which, when discussed, took on practical shape and form; and his comments and asides were memorable as was Harry Lawrence's face of horror when, discussing some customer's wish to be refunded for repairs which had subsequently gone wrong, he stated, 'Not one penny does he get back, Harry; we are here to make gold, not guns!' The accountants' annual visit to the Long Room was always highly entertaining as Uncle Hugh could usually find the one flaw in their figures which would alter everything which had been prepared. Certainly never before had they been to a board meeting where the chairman and three directors would drink whisky throughout the proceedings and they were told, 'You can have some afterwards!' In fact, all of us in Purdey's and the Purdey family can be thankful for that evening in the bar of White's Club when Hugh Sherwood decided to buy the Purdey shares to help out Tom.

On the death of Lord Sherwood, his brother, Sir Victor Seely, became chairman. The situation, due to the loss of craftsmen, was grave. Only two qualified stockers remained — Harold Delay, the brother of Ben Delay, and William O'Brien — and had either of these craftsmen slipped and broken an arm or even a finger at that time, Purdey's would have come to a

halt as the actioners would have known that the actions they were building could not have stocks fitted, and the finishing shop would have had no work; furthermore, the apprentices would not have been trained. For the best part of a year we lived with this fear while inflation increased and with it the natural demand for higher wages. Customers threatened to sue us for breach of contract if their guns were not delivered on time and one Italian industrialist was so incensed at the delay in finishing his over-and-under that he sent a cable to say that he was flying to London in his own jet to confront us with his lawyer. On the day appointed we waited in the Long Room dreading the inevitable scene, but at 11am a telephone message was received to the effect that the fog was so bad at Milan airport that the plane could not take off, the customer had returned home and wished to cancel the order, and we were to refund his deposit immediately. There were sighs of relief all round and thanks to a kindly Providence were offered up.

Seeing the difficulties under which the firm was labouring, more men left and so the delays in delivery increased. We again wrote to all our gun customers explaining the position, and apologising for further delays in completing their orders. We realised that the loss on each gun would increase with every extra per cent of inflation, on the books we were responsible for £100,000 worth of deposits all of which were repayable on demand, and almost all this money had been employed in the running of the business. The lease of South Audley Street and the factory were our only assets and they were not much good as they would come up for renewal during the next few years.

The majority of American customers reacted in a most generous and understanding manner, over 75 per cent of them agreeing to pay whatever surcharge might be necessary when their guns were finished, and we agreed to show them the costings supporting these extra charges. They told us not to hurry the work as they fully understood our position. The English and European customers were not nearly so accommodating.

Slowly the position improved. From a drop in production to only 49 guns in 1970, we built 55 the following year and 63 in 1972. Trained men who had left returned when it looked sure that things were improving, and we got in touch with others who subsequently returned to the new factory. A publicity campaign was set afloat as a result of which several boys applied for apprenticeships, and we started to settle down once more although there were to be another couple of years of appalling inflation to contend with.

From now on an open-ended escalation clause had to be included in all orders for new guns. With a delivery of up to four years and increasing inflation, it was impossible to foresee how high wages would rise in that

Sir Victor Seely Bart; director 1946, chairman 1970

period, so it was not possible to estimate the amount the customer might have to pay when the guns were completed. This clause was generally accepted, but some customers quibbled at the percentages which had to be added to their final bills. The only other remedy was to raise the price of the gun to a level which we considered would cover most of this eventual increase. A rise of this amount would almost certainly have stopped all future orders stone dead, so smaller increases were still decided upon in a vague hope that the government would eventually take positive steps to deal with this awful menace which was threatening our business and livelihood. Deposits on guns were fixed at one-third of the quoted price on ordering, another third when the gun was actioned, and the remainder on completion. Only thus could we help finance the building of guns over this long period.

In June 1971, Sir Victor Seely had reached the age of seventy-one and decided to resign the chairmanship to make way for a younger man, though he remained on the board as adviser of finance until December 1972. In 1971 therefore I took over the chairmanship, together with yet another overall increase in prices of everything from new guns to repairs, together with a revised pension scheme for all employees.

Once again the amounts owed to the firm by customers had increased significantly and a major drive on these outstanding debts was undertaken, as every pound was of prime importance in the circumstances in which we found ourselves. There was obviously no question of dividends being paid as there simply was no money available; all our resources had to be channelled to the financing of wages and pensions in order to keep the trained men while they completed the back-log of guns at all possible speed.

The daily ritual of going downstairs to the Long Room was an ordeal which had to be faced with courage and resignation as letter after letter hurled abuse at the firm over the continual delay in finishing new guns. As so few guns were being completed, the money supply virtually dried up and the envelopes containing cheques were eagerly examined and the value of their contents carefully counted until, with a sigh of relief, we reached the figure which ensured that there was enough money to pay the weekly wage bill. Certain telephone messages from the United States were particularly unpleasant with accusations of inefficiency, bad faith, stalling and general incompetence, usually delivered in a loud, rasping yell. As there was no short-cut available we just had to sit there and 'take it'.

Some letters were very difficult to answer. 'Dear Honourable Beaumont,' one furious American industrialist began adding, 'If you are so damned honourable why don't you stand by your word and deliver my gun on time?' I could not think of a reply to that one, but answering the one sent by an even angrier gentleman who swore that if he did not receive his gun within the next month he would sue me and 'take away all your ill-earned profits' was easy; there were no profits to take, and I reminded him of that fact. So it went on, each customer having to be placated, and assured that we would not speed up the delivery at the expense of quality, whatever action they took against us. Slowly and surely the position improved and gradually the horrifying back-log of under-priced guns was overcome and we could look ahead to once again building guns at a profit.

At the worst time, the over-and-under guns ordered in 1969 at a fixed price of £1,500 were taking five years to complete and were costing over £2,500 to build; and in some cases side-by-side guns were in a similar position. We determined that we would never again be caught in that spiral, so in 1971 we raised the price of the over-and-under gun to £5,000, to £7,500 in 1976 and to £15,000 in 1978. The side-by-side gun, which had stood at £2,000 in 1971, increased in yearly jumps to £4,500 in 1976 — a rise of 280 per cent in six years — and £7,500 in 1978. Inflation during those years continued at an average yearly rate of 25 per cent, though in 1974 it stood at 32 per cent. Fortunately, customers continued to order guns of this quality as a hedge against inflation and one customer, on receiving his new guns which had been built specially for him over a period of five years, took them straight to an auctioneer and sold them at current cost. Delivery dates had now gone from 3½ to 4 years so, taking advantage of the increase in prices during that time, he made a handsome profit.

Lawrence Salter on his selling trips to America had to contend with a great deal of criticism. The fear that our quality had been allowed to drop was actively encouraged by some interested parties in the gun world who saw obvious advantages for themselves if this could be proved. Lawrence Salter handled all these matters with great patience and skill and, with the constant support of his uncle, Harry Lawrence, who was still in close contact with his old American friends, pulled us through. In all this we were greatly helped by the support of Mr Carron Greig, chairman of Clarkson's Shipping who became a director in November 1972. In 1973 Mr Heath's confrontation with the miners caused further alarm when the electricity supply to the factory was cut to a three-day week. Once again the old Victorian gas brackets were brought out of store and Dan O'Brien, who had fitted them to the gas connections in 1947, did the same now, their light and heat enabling the men to continue normal working hours.

By 1976 the worst was over. The back-log of guns had been worked off, the delivery time had been reduced and we were able to face our customers with greater confidence. The resources of the firm had been improved, and when in 1979 a freehold factory site was available we were able to afford the site and build to our own specifications, the first time the firm has owned a freehold factory of its own. This occurred just as the rent of our old factory was due to be increased by 250 per cent.

We even absorbed without too much trouble the serious damage caused by an IRA bomb which, one evening in October 1975, exploded in a restaurant opposite the shop in South Audley Street. The blast destroyed every pane of glass in that face of the building, and blew the ornamental railings into the rooms of the Purdey flats. However, 'it is an ill wind', for as the plate-glass windows over the secretary's office collapsed, they took with them a panel of wood behind which was revealed all the Purdeys' private accounts and the firm's balance sheets covering the period 1880–1900, the existence of which had hitherto been unknown.

The previous year Purdey's had managed to bring back into the business No 84 Mount Street and, having knocked through the intervening wall between the front shop and this small premises, my wife opened a shop which caters for the wives of gun customers who can no longer bear the boredom of their husbands talking about guns in the Long Room. Shooting clothes designed by Mrs Beaumont for men and women, stockings, shooting garters, hats, caps, mittens, accessories for shooting and presents are sold and the shop has proved so successful that pressure had been brought to bear on the gun side to move this enterprise into the front shop; this suggestion has been firmly resisted.

In 1977, my cousin Nigel Beaumont joined the firm as an actioner's apprentice under Ben Delay. Brought up in Rhodesia, but educated at Shrewsbury, he obtained an Honours Degree at Manchester University and I met him for the first time at my uncle Ralph Beaumont's funeral in Wales. Nigel was a considerable artist and was fed up with 'only working with his brain'; he wanted to use his hands as well. Since then, his actions, barrels and stocks have gone out on both new and reconditioned guns and, after Harry Lawrence, he is the only man who can make a Purdey gun through all the processes, In 1981 he was sent out to Columbus, Ohio, on his own, to address the Ruffed Grouse Society on Purdey guns at their Annual Dinner. We thought there would be approximately 25 members present and told him not to worry; 460 members and their wives attended, so he was literally thrown in at the deep end. He acquitted himself with distinction and has since been to America on full selling trips. He was made a director in 1982 and with his gunmaking skills and organising ability he is, with the help of Lawrence Salter, maintaining the continuity of the firm.

With the firm's finances once more on a firm basis, the years from 1977 seem to be marked by celebrations rather than crises. In 1980 we celebrated the 100th Anniversary of the building of the front shop. This coincided with the Lawrence family's centenary of responsibility for building the guns, and in 1984 we celebrate One Hundred Years of the building of the Purdey-Beesley action. And on two Royal occasions — the Silver Jubilee of 1977 and the wedding of the Prince of Wales and Lady Diana Spencer in 1981 — the Purdey decorations have been displayed in all their glory.

The inlay on the Prince of Wales' stocks, showing the Prince of Wales' insignia beneath the Duke of York's (King George VI) insignia

   Not that things aren't a little more difficult now. On both occasions the full force of the bureaucracy of the Westminster City Council was brought to bear so that no fewer than five different permits had to be acquired before the decoration could be raised. Having satisfied every regulation concerning safety, lighting and stress, a young lady inspector came round to make sure that everything was in order. She looked up and saw Peter Stevenson fixing the shields to the railings on the first-floor balcony. 'Who is that?', she asked. It was explained that Peter was one of our employees helping with the decorations. 'If you use your own work force you require another permit', she said. And so it was to be! But, safely in place, the illumination drew crowds, and we received several letters from tourists and local residents expressing admiration and gratitude. One taxi driver came into the shop to tell us he admired the display so much that he brought his fares by way of Mount Street and South Audley Street when returning them home after dark so that they could see Purdey's; sometimes making the round trip rather expensive. We must now wait until the year

The present Chairman's insignia

2002 before the illumination will again be on display — to celebrate the Golden Jubilee of Her Majesty the Queen.

Meanwhile, in July 1981, seventy-four employees of Purdey's toasted the wedding of the Prince of Wales and Lady Diana Spencer in fourteen magnums of champagne in the new factory, and later a dinner party was held in the Long Room for all employees who had been with the firm for more than forty years. Sixteen qualified for invitations, fourteen attended; and thirty-three in all, wives included, sat down to dinner. If all those invited had been able to attend there would have been 910 years of service in the room, Harry Lawrence and Eugene Warner contributing 130 between them. Jok Purdey was present and so from the year before Bonaparte was beaten at Waterloo to two years after Galtieri was beaten in the Falkland Islands, the Purdey family have continued to build guns. Not many, it is true, as Serial No 1 was made in 1814 and Serial No 28,600 in 1983 — an average of only 170 guns, rifles and duelling pistols a year over those 168 years. But let us hope that the coming years will continue this happy record.

An early Purdey gun case label

APPENDIX 1

Deed of Agreement between James the Younger and his sons, James and Athol, changing the name of the firm to Purdey and Sons in 1877

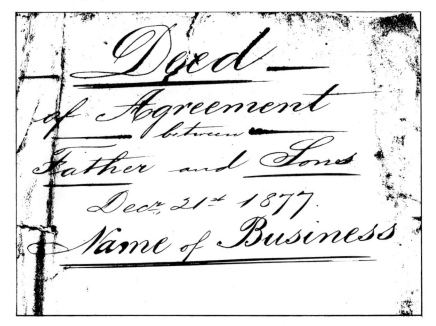

It is hereby understood and agreed **Between** James Purdey of No 314½ Oxford Street Gun Maker and his sons James Purdey the younger and Athol Stuart Purdey as follows

The said James Purdey having in contemplation certain arrangements which he may hereafter make for admitting his said Two sons to a share in his business of a Gun Manufacture carried on by him at 314½ Oxford Street aforesaid has, with a view to that possibility, and with a desire to benefit his said sons by familiarizing his Business customers at once with the step so contemplated, decided upon adding his sons names to his own by changing the name of James Purdey in which the said Business has hitherto been carried on to "Purdey and Sons" but on the distinct understanding and agreement with his said sons that the step so taken by him and the change in name under which the said Business is carried on shall only be for so long as the said James Purdey shall think fit And that the adding his sons names to his own is not to be taken as conferring any partnership rights powers or interests upon them his said sons in any way whatever

Now therefore in pursuance of the premises and for the considerations aforesaid It is hereby understood and agreed by and between the said James Purdey and the said James Purdey the younger and Athol Stuart Purdey That in the event of their Father the said James Purdey adding their names to his own in his Business description by the title of "Purdey and Sons", and having that title inscribed in the business premises or on Bill Heads or in other ways as the said James Purdey shall think proper, that the so doing by the said James Purdey shall not be taken or considered as giving to or conferring upon them any partnership rights or interests or any interest whatever in the business but that the said Business notwithstanding the use of their said names in manner aforesaid shall be that of the said James Purdey and shall belong to him alone And each of them the said James Purdey the younger and Athol Stuart Purdey doth hereby agree with their father the said James Purdey that he will not at any time or times hereafter under any circumstances whatsoever set up or lay claim to any share or interest in the Business of the said James Purdey as a Partner or otherwise until such time as the said James Purdey shall voluntarily make over to him any portion of the business as he may in his discretion deem advisable and proper And moreover that neither of them the said James Purdey the younger and Athol Stuart Purdey shall or will directly or indirectly do or say anything to indicate or lead any customer or other person to infer that either of them has any share or interest in the said Business nor do or commit any act deed or thing which shall in any way be an annoyance or embarrassment to the said James Purdey or contrary to the Spirit and intention of this agreement or the understanding come to between him and his said sons as aforesaid And lastly it is understood that the said James Purdey shall notwithstanding anything herein contained have it in his power at any time hereafter of his own free will, and without calling for a requiring the consent or concurrence of any other person to take down and remove the names of his said sons or either of them from his business title, and to restore his own name as it originally stood or make any other alterations therein as he shall think proper.

Dated this twenty first day of December One thousand eight hundred and seventy seven.

Witness to the signatures of James Purdey, James Purdey Jr. and Athol Stuart Purdey

George H. Lubbock,
314½ Oxford Street W.

James Purdey —
James Purdey Junr
Athol Stuart Purdey

APPENDIX 2

Alterations to the Woodward/Purdey over-and-under gun by Ernest and Harry Lawrence after 1948

1. Reduction in width of 12-bore 2¾in case action ie from 1.4in to 1.370in (¹/₁₆in) making the heavy pigeon action the same size as ordinary game gun (2½in case) without losing any strength using a better quality steel and reducing the stress of explosion by the increased strength of the walls of the action over a smaller area.

2. Firing pins redesigned to give maximum *BLOW* to centre of firing pin ie facing tumblers so that the blow is square to firing pin, and different shape to increase strength against breakage and to provide return springs.

3. Change of lock. From the frail swivel type to a roller on cam face, and more advanced type of lock in preparation.

4. The fore-end no longer to be made of mild steel but of KE805 steel far greater strength when opening gun ie the pressures of cocking locks, compressing ejector mainsprings and final arresting movement of barrels on action.

5. A rounded back to reduce weight and to keep a fine line into stock, use of a hidden bolster to keep line of detonating simple and smooth without losing any strength and making for a very bold line around the face of the action.

6. Use of Purdey rounded type of strap to prevent cracking of stock.

7. Use of Purdey type of safety slide for ease of operating safety and minimum maintenance.

APPENDIX 2

8 A Purdey shaped lever for looks and keeping this the same shape in all Purdey guns.

9 The 20-bore much the same size as the old Woodward type possibly just a little smaller in width but all other modifications as 12-bore.

10 28 and .410-bore   Complete redesigned actions with all of the improvements of 12-bore but with 12½ tons back pressure on face.

APPENDIX 3

The Numbering of Purdey Guns 1814–1983

All the guns, rifles, pistols, duelling-pistols and over-and-under guns built by Purdey's since 1814, whether flintlocks, percussion, pin-fire, centre-fire or of the Purdey/Beesley self-opening action, including those of B,C,D and E Quality, have been numbered in sequence of building from Gun No 1 in 1814 to No 28600 in December 1983.

A rough guide to the numbers of all the guns built since the introduction of the self-opening action in 1884 is given below.

Number of Gun	*Month and Year of Completion*
12,000	October 1884
13,000	August 1885
14,000	September 1891
15,000	July 1894
16,000	October 1897
17,000	November 1900
18,000	December 1903
19,000	May 1907
20,000	May 1911
21,000	July 1914
22,000	May 1921
23,000	June 1926
24,000	January 1930
25,000	June 1935
26,000	November 1947
27,000	June 1962
28,000	February 1974
28,600	December 1983

Nos 25000/1 are the pair of miniature guns built for King George V's Silver Jubilee in 1935

Between June 1925 and December 1983, 266 over-and-under shotguns have been built

ACKNOWLEDGEMENTS

I am deeply grateful to Her Majesty the Queen, Queen Elizabeth the Queen Mother, HRH Princess Alice, Duchess of Gloucester and HRH the Duke of Kent for permission to publish certain letters and photographs; to HRH the Prince of Wales and the Earl of Snowdon for permission to publish photographs of guns in their possession.

To my wife Lavinia for insisting that this book should be written; to all the staff at Purdey's and especially to Harry Lawrence, Eugene Warner and Graham Tollett for their reminiscences and help in researching this book. To Patricia Garnett, Annabel Buchan, Jean Hodgson and John Graham for their excellent typing of the various manuscripts; Miss Fairhurst for all her work in tracing the Purdey family wills; Tony Hutchings for being so patient over the selection and reproduction of photographs and portraits and, of course, to all the members of the Purdey family who provided anecdotes and family information. With special thanks to John Purdey for allowing me to reproduce Sefton Purdey's sketches. Also my grateful thanks to Harold Balfour for giving me so much information about the war years of 1914-18, and to Keith Hutchinson, the great-great-great nephew of Thomas Keck Hutchinson for his help with early records.

INDEX

Page numbers in italic indicate illustrations   See maiden name (pp8-9) for wives of Purdey family

Abercorn, Duchess of, 76
Abercorn, Duke of, 46
Abercrombie & Fitch (New York), 146, 206, 212
Acton, Harry (barrelmaker), *193*, 201
Afghanistan, King of, 207
Alba, Duke of, 72, 74
Albemarle, 4th Earl of, 22
Albert, HRH Prince Consort, 27, 29
Albert Victor, HRH Prince, 47
Alexandra, Queen Consort, 38, 92-3, 102
Alfonso XIII of Spain, 101-2, *101*, *110*, 123, *124*, 138, 164, *164-5*
Amalgamated Engineering Union, 176, 184, 204
Annesley, Col the Hon H., 37
Antwerp Pigeon Club, 61
Apprenticeship system, 190-2, 196, 203
Archer, Fred (jockey), 76
Astor, W. W., 76
Audley House: built, 51-4, *52*, *54*; decoration for, 83, 232; war damage, 176-8, *177*
Australian market, 48, 76, 215
Austria, Prince Imperial of, 47

Bad debts, 16, 20, 25-6, 28, 35, 74, *82*, 82-3, 138-9, 155
Bailey, Abe, 76
Balfour, 1st Lord of Inchrye, 126-7, *127*, 128, 134, 179, *184*; director, 205, 206-7, 209; retires, 218
Barcelona, HRH The Count of, 163
Barr-Smith, Thomas, 215
Battenberg, Prince Henry of, 74
Battenberg, Prince Maurice of, 110
Bavaria, Crown Prince of, 19

Beagle, HMS, guns and accessories for, 23, *24*
Beaumont, Major the Hon Edward, 122
Beaumont, Nigel (director), 232
Beaumont, Hon Richard: joins Purdey's, 212, 215; to Australia, 215; to Spain concerning General Franco accident, 222; USA and Japan, 222-3; chairman of Purdey's, 230
Beaumont, Hon Mrs Richard, opens accessory shop, 232
Beaumont, Violet, Lady (of Carlton Towers), 76
Beaumont Wentworth, 47
Beckford, William, 21
Bedell-Smith, General, 180, 187
Bedford, Duke of, 46
Beesley, Frederick (gunmaker), 60, 65-6, 146
Beit, Alfred, 76
Bentinck, Lord Henry, 21, 28
Birley, Sir Oswald RA, 79
Bonaparte, Prince Jerome Napoleon, 27
Boss, Thomas (gunmaker), 15, 103-4
Boswell, Sir James, 21
Bourbon Parma, HRH Prince Sixte de, 134
Brooke, Sir Victor of Colebrooke, 39, 40
Brookeborough, Viscount, 110
Brougham, Henry, 13
Brougham, James (brother), 13, 35
Buccleuch and Queensberry, Duke of, 74, 121-2
Buenos Aires Pigeon Club, 102

Cardigan, Countess of, 46
Cardigan, 9th Earl of, 26

Carol, HM King of Rumania, 134, 138, 168
Cartridge Turnover Machine, 30
Cartridges: demand for, 102, 104, 122, 144, 180-1; loading, 61, 63, 181
Cercle des Patineurs, 45, 61, 62
Chatham, 2nd Earl of, 13
Churchill, Lady Randolph, 74
Churchill, Winston, 80
Clark, Charles (head shopman), 92
Clark, Rev Nasson, 74
Clarke, Sir Rupert, 76
Clinton, Lord, 28
Cobbold, Capt, later Lt Col Ivan, *163*, 179, 185, 187
Cobbold, Capt Ralph, 224
Coburg, Duke of, 47
Collins, (Major) Sir Arthur (director), 215
Collis, James (gunmaker), 20
Conroy, Sir John, 19
Constantine, Grand Duke of Russia, 29
Corbett, Mr Vincent, 139
Coronation celebration, 83, 112, 170
Coventry, Aubrey, 46
Credit service introduced, 218
Crimped Turnover Cartridge, 217
Cumberland, Duke of, 18-19

Dalmeny, Lord, 122
Damascus barrels, 195
Darwin, Charles, 23, *24*
Davidson, Tommy (actioner), 202
de Grey, Earl (later 2nd Marquess of Ripon), 30, 36, 40, 48, 55, 61, 78-9, *78*, *97*
de Grey (later Lord Walsingham), 30, 36, 40, 61, 70, *77*, *97*

245

INDEX

Delay, Ben (actioner), 183, 232
Delay, Harold (stocker), 227
Denmark, Prince Charles of, 93
Diamond Jubilee (1897), 83
D'Orsay, Count, 27
D'Oyley Carte, L., 20
'Drop-down' gun, 29
Duncombe, Capt W., 46
Dunhill, Sir Thomas, 173
Durham, Earl of, 110
Dykes, Sir Percival, 36

Eastcote, South Harrow, shooting grounds, 145, 170
Edinburgh, HRH The Duke of, 207; grants warrant, 219
Edmonstone, Lt (later Cdr) 'Eddie' RN, 131, 141
Edward VII, King: grants warrant, 30; first Purdey gun, 38; India, 45-6; succeeds, 95, *107*; death, 110, *110*
Edward VIII, King (Duke of Windsor): first gun, 93, *94*; France, 130; correspondence, 148; abdicates, 170; later letter, 212, *213*
'Egg bomb', 162
Eisenhower, General Dwight D., 180
Eley, Charles, 91
Elizabeth II, HRH The Queen: coronation, 83; grants warrant, 219
Elstree shooting ground, 70
Evans, William (gunmaker), 59-60
'Express' rifle, 48

Fairbanks, Sir Douglas, 215, *216*
Farouk, King of Egypt, 171
Field, Mabel (wife of Athol Purdey), 84-6, *85, 114-15*; birth of sons, 86; illness, 86, *89*, 89-91; sons' education, 99, 116; war work, 128; family relationships, 124, 142, 156; death, 174
Fife, Earl of, 37, 66, 70
Fincastle, Lord, VC, 93
Fitzroy, Capt Robert RN, 23, *24*
'Flash' gun, 162
Fluid-steel barrels, 195
Forsyth, Rev Dr (gunmaker), 13, 19, 35
Fortescue, 1st Earl, 16
Foxholes, shooting ground, 31
Franco, General, 222
Franz Ferdinand of Austria, Archduke, 61, 76, 78
Fraser, Admiral of the Fleet, Sir Bruce, 131, 170, 185, *186*

Fullalove, Alfred (actioner), 132, 156

Gadsby, Chris (factory manager), 178
Galitzine, Prince D.B., 80
Gambier, William, 21
'Gamekeepers' guns', 20
George V, King: 30, 61, 77, *111*; first Purdey gun, 47; coronation, 112; France, 130; illness and wooden gun, 147-8; presented with miniature guns, 158, *159*; death, 168
George VI, King: first 16-bore gun, 110; France, 130; correspondence and new guns, 148, *150-1*; coronation, 170; orders guns for Prince Philip, 207
Gloucester, HRH The Duke of (1825), 18
Gloucester, HRH The Duke of, 148-9, *149*
Goatley, W.A. (solicitor), 16, 25
Goldsmid, Lionel (first entry in ledgers, 1818), 16
Gore, Hugh, 77-8
Grace, Dr W.G., 75-6, *75*
Grant, Stephen (gunmaker), 40
Greig, Carron (director), 231
Grey, William (gunmaker), 15
Grey, Sir Edward (of Fallodon), 74
Grubb, Messrs (Philadelphia), 45
Guinard et Cie, 121, 134, 145
Gunmakers, Worshipful Company of, 27, 39, 44-5
Gustav V of Sweden, 110

Haig, Capt Douglas, later Field Marshal Earl, 74
Harley, Sir Stanley, 179
Harvey, Alfred (barrelmaker), 204
Haullier, Monsieur (French inventor), 19
Haverson, Julia (2nd wife of James Purdey the Younger), 41, *43*, 90-1, *92*, 98-9, 109
Hawker, Colonel, 14
Heathcote, Frank, 37
Heathcote, Capt J. Edwards, 93
Herbert, W.H. (solicitor), 83
Hill, William (barrelmaker), 202
Hiring out of guns, 21, 46
Hopetoun, 5th Earl of, 18, 26
Howard de Walden, Lady, 36
Howard, Len (side-by-side actioner), 160

Howes, Edward (secretary of Purdey's), 168, 180
Hunt, Ken (gun engraver), 221
Huntingfield, Lord, 30, *97*
Hutchinson, Thomas Keck (gunmaker), 11-13

Ilchester, Earl of, 21
Imperial Metal Industries (IMI) Ltd, 217
Iniwra, Nawab of, 46
Isabella II, HM Queen of Spain, 61
Ismay, Bruce J., 77
Italy, HM King of (1861), 36
Italy, HM King of (1890), 74

Jaffrey, H., 46
Jam Sahib, Maharajah of Nawenager, 168
Japan, trade with, 222-3
Jeffries, Lincoln, 48
Johnstone, Sir F., 46
Jones, W.R. (1st alteration to safety, 1827), 22
Jung Bahador, Nepalese Ambassador, 29

Kaiser Wilhelm II of Germany, 110, *110*
Kell, Harry (gun engraver), 204
Kennedy, Lord (Aberdeen Rifle Club), 20
Kent, HRH The Duke of: correspondence, 148, *153-4*; death, 181
Kimball, Sir Marcus MP, 225
Kinloch, Harriet Patricia (1st wife of James Purdey IV), 126, 129, 131
Kinnoull, Earl of, 40
Kosygin, Alexei, 81
Khrushchev, Nikita, 81

La Boyteaux, Mary Stuart (3rd wife of James Purdey IV), 154-5, *155, 172*, 178, 214
Lancaster, Charles (barrelmaker), 15, 19, 20, 29
Landseer, Sir Edwin, 30
Lane, Charles (finisher), 175
Lane, Rev E.M. of Yarmouth, 18
Lang, Joseph (gunmaker), 19, 20
Lascelles, Viscount, 137-8, *138*
Lawrence, C. Harry (gunmaker, director): joins Purdey's, 122; joins army, 132; Queen's Dolls' House guns, 139; factory reductions (1932), 156; Jubilee guns (1935), 158, *159*; over and under guns, 159-61; clay-pigeon rocket,

246

INDEX

162; war, 171-2, 176-89; MBE, 189; post war, 207ff; 'Long Room' committee, 225; Russia (1975), 80
Lawrence, Ernest (gunmaker), 158
Lawrence, Ernest Charles (gunmaker, factory manager), *157*, 198-9, *199;* factory reductions, 156-7, 165; war production, 171; death, 218
Lawrence, Ralph: cartridge loader, 145; war production, 181; assistant proof master and death, 218
Lawrence, Thomas (factory manager), 198
Leeper, William (cartridge loader), 144-5
Le Faucheux (French gunmaker), 29
Leinegen, HRH The Prince of, 18
Lewis, Henry (head stocker), 35
Limited company status, 142, 143
Londonderry, 6th Marquess of, 74
Long Room, 53-4, *54*, 75, 147-8,*147*, 156, *166-7*, 171, 178, 180, 221-2
Lowe, Leonard (assistant secretary), 139

Macmillan, Rt Hon Harold, 178-9
Malmesbury, 3rd Earl, 28, 39
Manton, Joseph (gunmaker), 13-15, 19
Manton's of Calcutta, 40
Manzenado, Marques of, 163
Mary, Queen Consort: rifle for Prince George, 92-3; Dolls' House, 139, *140*; letter re Duke of Kent, 181, *182*
Maulay Hafid, King of Morocco, 145
Menzies, Sir Stewart, 179
Middleton, 6th Lord, 18, 26
Middleton's of Calcutta, 19
Miles, 'Mickey' (finisher), 139
Miller-Mundy, Major Godfrey: shares in Purdey's, 168; row with Tom, 170; sells, 185
Miniature guns, 139, *140*, 158, *159*
Moira, Lord (Master General of Ordnance), 13
Montreuil, Count de, 47
Morgan, William (shooting coach), 145-6, 162, 170
Mountbatten, Lord and Lady Louis, 137
Murray, Capt Verginius, 25

Muscat, Imam of, 26
Mysore, Maharajah of, 162

Newcastle, 4th Duke of, 26
Nicholas I of Russia, 29
Nicholas II of Russia, 80
Nickerson, Sir Joseph, 161-2
Nizam, His Exalted Highness, of the Deccan, 46
Nobbs, William (actioner, inventor), 103, 202
Northwick, Lord, 36
Nugent, Colonel Tim, 179

Obolensky, Prince, 138
O'Brien, Dan (Purdey craftsman), 201
O'Brien, William (stocker), 227
Oliver, Beatrice (2nd wife of James Purdey IV), 137, 142, 154
Oliver, F.S. (father): owner of shares, 142, 155; sells, 160
Orange, HRH The Prince of, 19, 27
Orleans, Infante Don Antonio, 74
Over-and-under gun, 37, 159-62

Paris, HRH Comtesse de, 76
Paris Exhibition (1878), 48
Payne-Gallwey, Sir Ralph, 63-5, *64*, 108, 193
Peel, Sir Robert, 21
Pellew, Hon P.J., 25
Pennell, Mr Cholmondeley, 40
Persia, Shah of, 207
Philippi, George, 134
Pickering, Rev Henry, 36
Pickersgill, Mr, 36
'Pin-fire' gun, 29
Portal, Sir Wyndam, later Lord, *169*; (owner of Purdey's), 168, 170-1, 178; sells, 183
Portugal, Crown Prince of, 74
Pound, Sir Dudley, 168
Powell, William L. (gunmaker), 95-8
Princess Royal, HRH The Countess of Harewood, *138*; letters to Mabel Purdey, *114-15*, 128; tea with Tom at Eton, 116; buys gun, 138
Prussia, HRH Prince Frederick, 168
Purdey Bolt, 30, 39
Purdey family genealogy, 8-9
Athol Stuart (1858-1939), *57*, *94*, *133*, *143*, *147*, *172*; boyhood, 33, 41-4; marriage and children, 84-6, *89*, 89-91; breach with Powell's,

95-8; law case with Boss, 103-4; war work, 122-3, 127-8; Queen's Dolls' House, 139, *140*; special gun for King George, 147; miniature guns for Silver Jubilee, 157-8; death, 173
Cecil Onslow (1865-1943), *119*; birth, 33; joins firm, 69-70; director, 143, 155, 179, 183; death, 185
James (b1737), 11, *12*
James 'The Founder' (1784-1863): early years, 11-13; with Manton, 13; with Rev Forsyth, 13; own shop, 14; acquires Manton's shop, 19; marriage and children, 19; Gunmakers' Company, 27; royal warrant, 30; death of wife and self, 33-4
James 'The Younger' (1828-1909), *81*, *92*; birth, 19; marriage and children, 32, 33, 38; second marriage, 41, *42*; Gunmakers' Company, 44; cheaper guns made, 48-51; builds in South Audley Street, 51-4; Beesley Patent, 65-6; alters Long Room, 72; royal patronage, 76-8; Diamond Jubilee celebrations, 83; death and tributes, 107-9
James III (1854-90): birth, 33; illness, 41-4, 69-74, *73*; death, 74
James IV (1891-1963), *147*; early years, 86, *89*, *100*, 99-100; army, 101, 112-13, *113*, 125-6, 128; marriage, 126, 129, 131; remarries, 137; joins Purdey's, 142; third marriage, 154, *155*, *172*; fiftieth birthday party, 178-80; recommends closing gunmaking, 206; death, 223
James Oliver Kinloch (Jok), Major (b1918), 44, 129, *172*
John (c1690), 11
Martha (1774-1853), 11, 33
Percy John (1869-1951), 33, *92*, 98, 99
Sefton (b1877): birth, 41; sketches, 55; Boer War, 86-7, 89-90
Tom (1897-1957), *147*, *157*, *169*, *184*; early years, 86, *87*, 112-13, 116, *119*; services, 119-20, *119*, 124-5, 126-31; joins Purdey's, 131; presentation guns for King George V, 157-8; preparation for war, 171-3; Ivan

247

INDEX

Cobbold buys shares, 185; Lord Sherwood buys shares, 189; post-war difficulties, 205-10; brighter prospects, 212; crimped cartridges, 217; retires, 218-19; death, 219-20

Reilly (gunmaker), 58-9
Remington-Wilson, R.H.R., 74, 108
Ribbentrop von, 165-6
Richards, Sir Gordon, 76
Richmond, Duke of, 70, 122
Richmond Watson, Mr (West London Shooting Grounds), 170
Rigby (gunmakers), 37
Ripon, 1st Marquess (Viceroy of India), 48
Rhodes, Rt Hon Cecil, 80
Robertson, J. (Boss & Co), 15, 103-4
Rocket-propelled pigeon, 162
Rose, Michael (West London Shooting Grounds), 170
Rothschild family, 30, 76
Royal warrants, 30, 48, 78, 110, 172, 219
Rutland, Duke of, 46

St Albans, Duke of, 82, *82*
St Leger, Captain, 21
Salter, Lawrence (managing director), 44; joins Purdey's, 184; gun apprentice, 209; visits USA, 224; joint managing director, 226
Savile, Lord, *105*
Sawdust cartridges, 39
Schakel, 'Gus' (stocker), 139
Scott Thompson, Captain, 25
Seely, 'Jack' (1st Lord Mottistone), 79-80
Seely, Major Sir Victor, 44; joins Purdey's, 205; policy, 206ff; chairman, 227; retires, 230
Sefton, Earl of, 61
Serbia, King and Queen of, 77, 81, 134, 145
Sherwood, 1st Baron (Sir Hugh Seely Bart), 80, *209, 229*; buys Purdey shares, 189; policy, 204ff; offers firm to author, 212; sole agency issue, 212; tries to reduce Purchase Tax, 214-15; death, 226-7
Shilston, T., 59
Shuttleworth, Frank, 56-8
Siam, King of, 36
Singh, Prince Duleep, 30, 46, 53, 55, 74, *96*
Single-trigger gun, 103
Skouras, Charles P., 217
Smith, Harry (shooting coach), 146
Solo (Java), Emperor of, 30
Somerville, Rev John, 22
Sopwith, Thomas, 47
Spencer of Northallerton (gunmaker), 22
Stanbury, Percy (shooting coach), 170
Stewart-Wortley, A., 75
Stonor, Sir Harry, 78, 122; death, 173-4
Swannell, Major David, 76

Tankerville, Earl of, 21
Teba, Conde de, 163-4
Teck, Duke of, 74
Thomas, Caroline (wife of James Purdey the Younger), 33, *34*, 38
Thompson & Poole, Messrs, 21
Tillson, George (head clerk), 143, 168
Tipping of servants, 38, 46
Tir au Pigeon (San Sebastian), 102
Tollett, Graham (clerk) 16, 143, 144, 171, 181
Top lever invention, 30
Trafford, Sir Humphrey de, 179
Trenchard, Lord (Marshal of the RAF), 126, 134, 137
Tryon, Vice Admiral Sir George RN, 47
Tryon, Hon Anthony, 226
Turkey, Sultan of, 61

van Lengerke & Detmond, 134
Vermin asphyxiator, 36
Vicars, Joseph (gunmaker), 13
Victoria, Queen: orders guns and accessories, 26-7, 29, 38, 47; Diamond Jubilee, 83; death, 93
Vivian, Hussey (1st Lord Swansea), 40

Walker-Okeover, Sir Ian, 145
Wallace, Sir Richard, 41
Warner, Eugene (accounts dept), 16, 132, 223, 234
Warren, Walter (engraver), 202
Wemyss, Captain, 26
Wernher, Julius, 76
West London Shooting Grounds, 146, 170
Westminster, 1st Duke of, 51
Wheatley, 'Sheriff' (factory manager), 192-3, 197, 201, 202
Whitworth, Sir Joseph (ironmaster), 195
Wiart, General Adrian Carton de, VC, 138, 179
Wilkes, Charles (factory viewer), 193
Winans, Walter, 36, *37*, 46, 53
Wire-mesh cartridge, 39
Wolff-Metternich, Count von, 120
Women workers, 178, 187
Wood, George (barrelmaker), 161
Woodward & Sons (gunmakers), 199-200; taken over, 160-1
Worshipful Company of Gunmakers, 27, 39, 44-5
'Wurtenberg, Roi de', 27